Springer Tracts in Advanced Robotics 91

Editors

For further volumes:
http://www.springer.com/series/5208

STAR (Springer Tracts in Advanced Robotics) has been promoted under the auspices of EURON (European Robotics Research Network)

João Filipe Ferreira · Jorge Dias

Probabilistic Approaches for Robotic Perception

Springer

João Filipe Ferreira
Instituto de Sistemas e Robotica
Departamento de Engenharia
Electrotécnica e Computadores
Pinhal de Marrocos, Pólo II
Universidade de Coimbra
3030-290 Coimbra
Portugal
jfilipe@isr.uc.pt

Jorge Dias
Instituto de Sistemas e Robotica
Departamento de Engenharia
Electrotécnica e Computadores
Pinhal de Marrocos, Pólo II
Universidade de Coimbra
3030-290 Coimbra
Portugal
jorge@isr.uc.pt

ISSN 1610-7438 ISSN 1610-742X (electronic)
ISBN 978-3-319-03289-4 ISBN 978-3-319-02006-8 (eBook)
DOI 10.1007/978-3-319-02006-8
Springer Cham Heidelberg New York Dordrecht London

Printed on acid-free paper

Springer is part of Springer Science+Business Media (www.springer.com)

This book is dedicated to the better part of me:
my wife, Anita, and my children, Luísa and Miguel

J.F.F.

This book is dedicated to all students and colleagues that with their infinite patience and perseverance have been companions in my research activities

J.D.

Preface

This book tries to address the following questions: How should the uncertainty and incompleteness inherent to sensing the environment be represented and modelled in a way that will increase the autonomy of a robot? How should a robotic system perceive, infer, decide and act efficiently? These are two of the challenging questions robotics community and robotic researchers have been facing. The design of more autonomous, more intelligent and adaptive artificial system is the context of this book, with a particular emphasis on probabilistic techniques and Bayesian inference.

Purpose

In this book we will show how probabilistic models and Bayesian inference can be used to develop and improve robotic systems to perform complex tasks in real world environments. In particular, recent advances on the topic will be presented, mainly from research activities on artificial robotic perception.

The book provides background on the Bayesian inference techniques that allow researchers and students to address the question of how different sensory modalities can be processed to converge to form a coherent and robust percept of the environment.

In this text, three major issues concerning probabilistic robotic perception modelling are addressed:

- Representation of three-dimensional space within a probabilistic framework;
- Hierarchical combination of Bayesian models and representations;
- Definition of decision and learning processes based on Bayesian programming and models.

Who This Book Is For

This book provides an introduction to the use of probabilistic tools to implement robotic perception, adding to it working examples and case studies. As such, it should be helpful to many different kinds of readers:

Researchers and robot developers
For seasoned researchers or professionals working in the field of robotics who wish to devise probabilistic solutions during the course of their work, the intuitions supporting the probabilistic modelling process described throughout the worked-out examples in the book can promote the necessary frame of mind, without alienating years of experience in artificial perception.

Students
The introductory background, intuitive explanations, detailed formalisation and sample code, downloadable from the companion website, will allow students to reduce the uncertainty and ambiguities of Bayesian modelling in robotic perception, work on more challenging projects, and ultimately contribute with cutting-edge research to the field.

Teachers
Probabilistic approaches to perception are in the forefront of robotics research, and are also becoming a common modelling language across many other research fields. Therefore, the availability of a textbook with a companion website packed with supplementary material, such as sample code complementing worked-out examples and assignments, is paramount to flatten the learning curve, consequently spurring students into becoming proficient and considering more ambitious tasks.

Hobbyists
Probabilistic modelling and robotic perception together are fun!

Motivations

For perceiving the environment our brain uses multiple sources of sensory information derived from several different modalities, including vision, touch and audition. The question of how information derived from these different sensory modalities converges in order to form a coherent and robust percept is central to develop processes of artificial perception. The combination and integration of multiple sources of sensory information is the key to robust perception, because no information processing system, neither biological nor artificial, is powerful enough to actively and accurately perceive under all possible conditions.

The development of robotics domain by the 1980s spurred the convergence of automation to autonomy, and the field of Robotics has consequently converged towards the field of artificial intelligence (AI). Since the end of that decade, the general public's imagination has been stimulated by high expectations on autonomy, where artificial intelligence and robotics try to solve difficult cognitive problems through algorithms developed from either philosophical and anthropological conjectures or incomplete notions of cognitive reasoning. Many of these developments do not unveil even a few of the processes through which biological organisms solve these same problems with a fraction of the energy and computational resources.

The tangible results of this research tendency were many robotic devices demonstrating good performance, but only under well-defined and constrained environments. The adaptability to different and more complex scenarios was very limited. As a reaction, an emergent trend has recently being surfacing, in which researchers look for biological inspiration to create a new generation of robots.

This new generation of robots address solutions for one of the major obstacles for reliable robotic autonomy — the extraction of useful information about the external environment from sensorial readings — in other words, artificial perception.

Contemporary robots and other cognitive artefacts are not fully capable of autonomously operating in complex real world environments. One of the major reasons for this failure is the lack of development of cognitive systems that are able to handle with the incomplete knowledge and uncertainty. The development of these artificial perception systems, focussed on multimodal and multisensory integration, will be necessary, but using computational/statistical models supported by observations of biological systems.

In this book, the topic of artificial perception is addressed. The application of Bayesian models and approaches is described in order to develop artificial cognitive systems that carry out complex tasks in real world environments. Throughout the book we will see how to apply models derived from research on cognitive systems in robotic frameworks. The Bayesian approach is used to model different levels of cognitive activities and coherently model these activities within the Bayesian framework. The Bayesian framework is clearly a multidisciplinary approach which has been used in an increasing number of scientific domains.

Prerequisites

The book uses probabilistic inference as a key tool for modelling robotic perception — it is therefore assumed that readers have a grasp of the basics in the fields of robotics, computer vision and signal processing.

On the other hand, the first chapter of the book provides a short summary of the main concepts used throughout the text, such as the fundamentals of probability theory and statistical inference, which provide the necessary background to allow the application of the Bayesian approach to robotics and artificial multimodal perception.

How This Book Is Best Used

The book shows how Bayesian models can be used to develop artificial cognitive systems. After a primer on probabilistic inference and an explanation on modelling perception as a process with sensing and action, the book shows how a Bayesian approach can be applied to artificial perception systems. The readers will have the opportunity to study and test several real examples on the application of these techniques in the robotics domain, including the implementation of artificial perception using sensory systems as diverse as vision, touch or audition.

In the first part of the book, the following scientific aspects will be addressed and illustrated using worked out examples:

- Fundamentals of Bayesian inference;
- Representation of three-dimensional space and sensor modelling within a probabilistic framework;
- Bayesian programming and modelling for robotic perception systems;
- Hierarchical combination of Bayesian models and representations;
- Decision and control processes for learning and motor control.

In the second part of the book, a couple of case studies applying these techniques to robotics are presented.

Many of the examples and applications take inspiration from models resulting from research on living systems and apply these models in the development of innovative and self-learning robotic systems. They are intended, not only to provide insight on the process of probabilistic model design in this context, but additionally to incentivise readers to come up with solutions of their own.

Finally, a website has been developed as a companion to this textbook, `http://media.isr.uc.pt/monograph`, which we strongly encourage to be used as a complement to the monograph itself, consisting of a user-friendly source of supplementary material, such as online interactive appendices, assignments, links to bibliographical references, tutorials, presentations and sample code.

Acknowledgements

The authors owe a substantial debt of gratitude to many people who, directly or indirectly, contributed to the inception of this book, namely those who have impressed their fingerprint on our work and influenced our thinking over the years.

First, we would like to thank all of our friends and colleagues from the BACS European Integrated Project ("Bayesian Approach to Cognitive Systems", FP6-IST-27140), our companions on the journey that motivated this book. In particular, we would like to express our gratitude to Pierre Bessière for his wisdom and friendship in countless conversations and modelling sessions on the whiteboard, and the inspiration he has been for a whole generation of practitioners of the Bayesian approach. Among this generation, we would like to extend a special "thank you" to Julien Diard – without him, the worked out examples would not have been as interesting, the chapter on Bayesian hierarchical constructs would not have been as rich, and the impressive interactive online appendix **"Probabilistic Approaches for Robotic Perception – Online Interactive Examples"**, written in collaboration for Wolfram's free CDF Player, would have been impossible. But the influence of this generation of brilliant minds in this book does not end here – we are deeply in debt for all the uplifting conversations, scientific discussions and support of "brothers in arms" such as Francis Colas, Manuel Yguel, and many, many others.

We are also deeply grateful to our Mobile Robotics Laboratory family, residing at the Institute of Systems and Robotics of the University of Coimbra. Many of the worked out examples and figures were reproduced from their excellent research work, which they kindly allowed us to use in our advantage. We would particularly like to thank Jorge Lobo, who might be thought as a kind of uncredited third author of this book, for his encouragement, insights and many, many years of friendship (and also the "Popular Science" issue credited in the introductory section!).

Special thanks are also due to Thomas Ditzinger, our senior editor, and to all the staff at Springer for their patience and support.

Finally, no-one deserves more gratitude than our family and close friends for their continued love, support and patience, as they missed our company (and sometimes our peace of mind) for endless evenings and weekends while we completed this book. We could have never performed this miracle without you!

Coimbra, Portugal,
August 26, 2013

João Filipe Ferreira
Jorge Dias

Notation

The notation throughout the main text is defined as new subjects are introduced in the course of the narration; in the authors' point of view, this makes absorbing new concepts much easier for the reader. However, the drawback of this approach is that the amount of scattered mathematical definitions presented herewith will perhaps be difficult to retain for readers who, while being familiar to the concepts, are unfamiliar with the corresponding notation.

With this in mind, the following list provides an overview of all important notation used throughout the text. It presents the respective shorthand definitions and also refers to the **page number where each of these were first introduced**, in case the reader needs a more in-depth and contextualised explanation.

A	Random variable, with few exceptions usually denoted with only one capitalised letter 8
A_i	The ith element of a set of functionally similar but usually independent random variables 13
${}^{t}A$	A random variable referred to a specific time instant t in a dynamic model 28
$[A = a]$ or a	Proposition (explicit and implicit versions) stating the instantiation of a random variable (unambiguous, either true or false) 21
A	By abuse of notation, an *unspecified* proposition stated regarding knowledge on a random variable with the same name 19

π_{idx}	Proposition pertaining to preliminary knowledge *idx* on (latent) implicit factors. Generally used to distinguish between different models (i.e. contexts) labelled by *idx*; if only one model exists, also used as π ... 23
δ_{idx}	Random variable denoting case *idx* of training data used for Bayesian learning ... 149
Δ_{idx}	Random variable summarising training data *idx* used for Bayesian learning ... 147
$P(A \mid B)$	Conditional probability: degree of plausibility assigned to proposition A given proposition B ... 20
$P(A)$	Degree of plausibility assigned to proposition A within a specific context – a shorthand for $P(A\|\pi)$; also, probability distribution: degree of plausibility assigned to unspecified proposition A stated regarding knowledge on random variable A and its respective measurable space – it is said that random variable A *follows* probability distribution $P(A)$ ... 21
$P(A = a)$	Probability: degree of plausibility assigned to the proposition of the occurrence a specific value a of random variable A ... 21
$E[f(A)]$	Shorthand notation for the expectation of a random variable A following distribution $P(A)$ ... 14
$E[A]$	Shorthand notation for the expected value (i.e. mean) of a random variable A following distribution $P(A)$ ... 14
$H(A = a)$	Information conveyed by the event of random variable A being assigned a specific value a ... 31
$H(A)$	Entropy of random variable A and respective probability distribution ... 32

Contents

List of Figures

List of Tables

List of Algorithms

Part I

Probabilistic Modelling for Robotic Perception

1

Fundamentals of Bayesian Inference

Probability theory is nothing but common sense reduced to calculation.
Laplace (1819)

Orthodox thinking, then and now, wants us to define probabilities only as physical frequencies, and deplores any other criterion as not "objective". Yet when confronted with a (literally) dirty, objective real problem, common sense overrides orthodox teaching and tells us that to make the most reliable inferences about the special case before us, we ought to take into account all the information that we have, whatever its nature.
"Where do we go from here?", in Maximum-Entropy and Bayesian Methods in Inverse Problems, *p. 21–58, Jaynes (1985)*

1.1 Introduction

Since the term *robot* (from the Czech or Polish words *robota*, meaning "labour", and *robotnik*, meaning "workman") was introduced in 1923 and the first steps towards real robotic systems were taken by the early-to-mid-1940s, expectations regarding Robotics have shifted from the development of automatic tools to aid or even replace humans in highly repetitive, simple, but physically demanding tasks, to the emergence of autonomous robots and vehicles, and finally to the development of service and social robots.

Along this journey from *automaticity* to *autonomy*, the field of robotics has unavoidably converged towards the field of artificial intelligence (AI) — one has but to notice that the word "autonomous" has its roots in the Greek for *self-willed*. As such, it has suffered the same fate as AI, losing some of its credibility and the power to stimulate the general public's imagination since the late 1980s. The reason for this is that both AI and robotics set expectations too high as for where autonomy was concerned, aiming for solving difficult cognitive problems through algorithms developed from either philosophical and anthropological conjectures or misconstrued notions of cognitive reasoning, without having yet unveiled even a few of the processes through which biological organisms, in fact, solve these same problems seemingly effortlessly.

One of the major obstacles for reliable robotic autonomy is the problem of the extraction of useful information about the external environment from sensory readings — in other words, *perception*. As of until recently, robots and computer-based devices had been in a state of sensory deprivation. This was in stark contrast with how humans and other natural organisms interact with their everyday environment, efficiently utilising a variety of sensors, such as visual, auditory, haptic, magnetic and odour sensing, just to mention

J.F. Ferreira and J. Dias: *Probabilistic Approaches for Robotic Perception*,
Springer Tracts in Advanced Robotics 91,
DOI: 10.1007/978-3-319-02006-8_1,

but a few, even for apparently simple tasks. On the other hand, even when taking into account how these organisms perform using each of these sensory sources individually, their efficiency and efficacy in doing so highly surpasses robotic performance in most cases. Regardless of the unfulfilled expectations, artificial perception has nonetheless witnessed major developments.

Introspection fools us into thinking that perception is deterministic and certain — indeed, as perceptual beings, we rarely question the veracity and accuracy of the understanding of the surrounding world resulting from our senses. As a matter of fact, in psychology the notion of "perception" has often been defined as "a *single unified* awareness derived from sensory processes while a stimulus is present". This is in accordance with what our common-sense tells us at first glance about what we perceive: "the room that I'm seeing has a chair, a table, and there's no doubt or uncertainty about that".

However, recent research on biological perception systems such as the perceptual pathways in our own brain are beginning to question this view. Doya, Ishii, Pouget, and Rao [2] demonstrate throughout their book how an alternative view, the idea that perception is a result of a probabilistic inference process, has surfaced in countless studies in neuroscience, suggesting that the human brain somehow represents and manipulates uncertain information, which can be described in terms of probability distributions.

The implications of these findings, in fact, surface against all of what our common-sense tells us, whenever we are confronted with scenarios which do not conform with what our brains are preprogrammed to accept, as *perceptual illusions* and *bistable percepts*. Nevertheless, the huge amount of natural scenarios which the human perceptual brain is able to cope with (apparently flawlessly) is absolutely astounding. On the contrary, it is particularly striking that robotic perception systems that attempt to tackle more generic and complex problems still cannot rival the performance of a three year-old child due to the lack of adaptive behaviour. Indeed, the amount of restrictions that have to be imposed for robotic perception systems to be able to deal with uncertainty using deterministic and ad hoc approaches is non-negligible (see Fig. 1.1).

When one of the authors of this text was in his teens[1], one day, while reasoning about Prolog programming and trying to replicate the ELIZA program by MIT professor Joseph Weizenbaum, at the end of that day thought "AI is a dead-end: as things stand, the if-then-else rationale and absolute, binary logic just don't make sense – they sure don't reflect more than a fraction of *how* we think!"

Aristotle, more or less two and a half thousand years ago, proposed the two famous *strong* syllogisms:

[1] Still dreaming of mobile and humanoid robots in distant planets after reading an issue of "Popular Science" left on his lap by a close friend (nowadays a long-time "partner in crime" in robotics research).

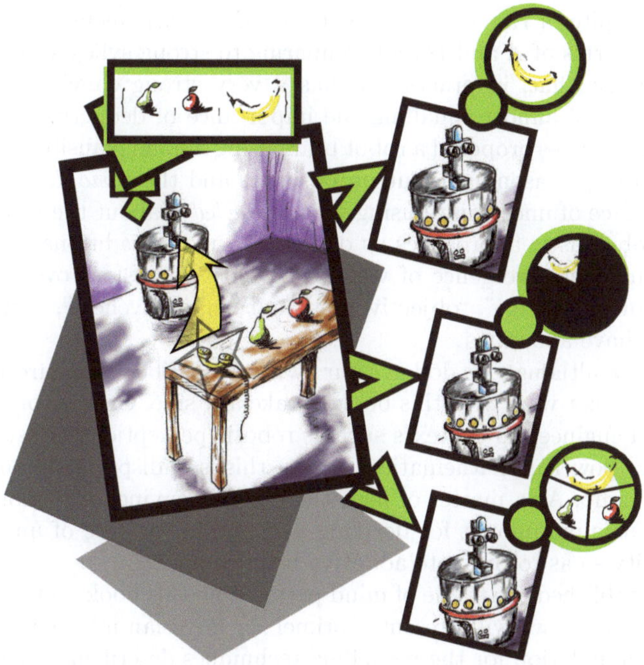

Fig. 1.1. Dealing with uncertainty in perception. Take the simple example depicted: a robotic perception system is faced with an object recognition task, for which the object being examined, in this example a handset, does not belong to any of the classes within the set the robot is prepared to identify, in this case an apple, a pear and a banana. The most common computational approaches used to solve such a perceptual problem, depicted as the outcomes stemming to the right of the main frame, would be [5]: deterministic solutions to reasoning about any realistic domain – enumerate (infeasible in general; top outcome) or ignore (within bounded error; middle outcome) exceptions; alternative solutions using uncertainty – summarise exceptions using numerical measures of uncertainty regarding propositions (bottom outcome).

$$\text{if } A \text{ is true, then } B \text{ is true} \Rightarrow \begin{cases} A \text{ is true, therefore } B \text{ is true,} \\ B \text{ is false, therefore } A \text{ is false.} \end{cases}$$

Jaynes [4] demonstrated that, as opposed to using these syllogisms, which are by definition formulated using *deductive reasoning* and absolute logic and, in most cases, impossible to apply since we do not have access to the right kind of information to use them, we are instead most proficient in applying a set of *weak* syllogisms, which are formulated using what is called *plausible reasoning*, of which the weakest would read

$$\begin{aligned} &\text{if } A \text{ is true, then } B \text{ becomes more plausible} \Rightarrow \\ &B \text{ is true, therefore } A \text{ becomes more plausible.} \end{aligned}$$

Indeed, in spite of the apparent weakness of this argument, when stated abstractly in terms of A and B and comparing to strong syllogisms, we have but to recognise that, in practice, it has a very strong convincing power, up to the point of almost equalling the importance of deductive reasoning. Consequently, Jaynes proposed a robot brain using human plausible reasoning as a black-box — reasoning about propositions and the *basic desiderata* for the robot in face of uncertainty using *prior knowledge* about the world. This vision of probabilistic approaches as opposed to more traditional views has led to a significant divergence of views, in which discussions over concepts such as "prior knowledge", "objectiveness" and "subjectiveness", "belief" and "frequency", have abounded.

In summary, although we do have our own personal beliefs regarding these matters, we do not wish with this book to take any side. On the contrary, we claim that in engineering contexts such as robotic perception Bayesian inference is a very powerful mathematical tool — this is undisputed among partisans of any camp. As a matter of fact, there is no denying, as demonstrated above, that it is paramount for deriving percepts in the face of uncertainty and ambiguity so as to promote adaptive behaviour.

Having established the frame of mind permeating this book, in the remainder of this chapter, we will present a primer on Bayesian inference that will serve as the foundation for the modelling techniques described herewith.

1.2 Statistical Inference and Sampling

1.2.1 Random Experiments, Events, Probabilities and Distributions

A *random experiment* is an experiment that satisfies the following conditions:

1. all possible distinct outcomes are known in advance;
2. in any particular trial, the outcome is not known in advance;
3. The experiment can be repeated under identical conditions

The *outcome space* Ω of such an experiment is the set of all its possible outcomes. Something that might or might not happen depending on the outcome of a random experiment is called an *event*. An event is defined as a subset of the outcome space Ω.

If all outcomes in Ω are taken to be equally likely[2], the *classical definition of probability* of an event A is stated as the number of outcomes in A, denoted as $M(A)$, divided by the total number of possible outcomes $M = M(\Omega)$:

$$P(A) = \frac{M(A)}{M}. \tag{1.1}$$

[2] Red alert: this is *prior knowledge in disguise*!

Example 1.1. **Tossing a coin**

Tossing a coin is a random experiment. Its outcome space is $\Omega = \{h, t\}$, where h and t mean "heads" and "tails", respectively.

An event "coin lands head", in this context, is represented as a single-element subset $A = \{h\}$. The probability of this event occurring is, according to equation (1.1), given by $P(A) = 1/2$.

Example 1.2. **Rolling a die**

Oddly enough considering the previous example, rolling a die is a random experiment with outcome space $\Omega = \{1, 2, 3, 4, 5, 6\}$. (Have we mentioned that probability theory is easy?)

The event "die shows an odd number" is represented as $A = \{1, 3, 5\}$. The probability of such an event occuring is, according to equation (1.1), given by $P(A) = 3/6 = 1/2$.

The assumption of "equally likely" limits the application of this concept — what if the coin or the die are not "fair"?

The question posed at the end of the examples presented above provided the motivation for the *frequentist definition of probability*, stated next. When the number of trials of a random experiment (remember: which may be repeated under identical conditions) is repeated indefinitely, the relative frequency of the occurrence of an event approaches a constant number

$$P(A) = \lim_{m \to \infty} \frac{m(A)}{m}, \tag{1.2}$$

where m is the total number of trials and $m(A)$ is the occurrence count of event A after m trials, the analogues of M and $M(A)$ in the classical definition, respectively. The law of large numbers states that this limit indeed exists, and in the frequentist perspective it is equivalent to the probability $P(A)$ of that event.

Example 1.3. **Tossing a coin (frequentist version)**

Imagine that we repeat the random experiment of tossing a coin $m = 1000$ times, while keeping the count $m(A)$ of the occurrences of event $A = h$.

Accepting that $m = 1000$ is big enough number of trials, the probability of $P(A)$ occurring, according to equation (1.2), seems to also be given by $P(A) = 1/2$. Indeed, Fig. 1.2 shows how one approximates this result as the number of trials increases.

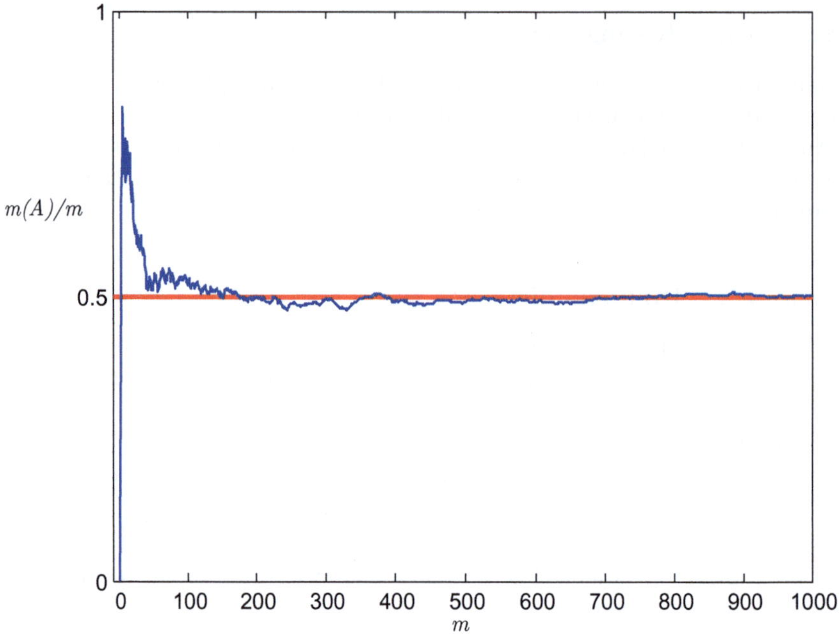

Fig. 1.2. Relative frequency of "heads" in a sequence of 1000 coin tosses.

This definition of probability gives us a way of testing the probability, for example, of unfair coins; however, since it is impossible to repeat the experiment an infinite number of times, we can only approximate the result of equation (1.2).

A *random variable* is a variable that is yet to be assigned a value; and, until the assignment is realised, we will remain *uncertain* about this value. For example, the height of a randomly selected person in a room is a random variable — we won't know its value until the person is selected. Note that we are not completely uncertain about most random variables; for example, we know that height will probably be within a certain range. Formally, random variables are an example of what is called a *measurable function*. Throughout this text, with very few exceptions, random variables will be denoted as capitalised letters, such as A.

Random variables are usually real-valued, but one can consider arbitrary types such as boolean values, complex numbers, vectors, matrices, and many others. Most commonly, one classifies random variables as either *discrete* or *continuous* (there are other classifications, but all beyond the scope of this text). A discrete random variable's *measurable space* (or, more simply, its space or support) is a countable set (e.g., the set of integer numbers), whereas a continuous random measurable space is an uncountable set (e.g., the set of real numbers).

Probability theory can be directly applied to random variables by assuming that *drawing* a specific value[3] for a given variable (or, in other words, *instantiating* it) is, in fact, an event, and can thus be assigned a probability of occurrence.

We can therefore formally define a random variable as a measurable function from a probability space to some measurable space. Consequently, discrete random variables *map outcomes* to values of countable sets, whereas continuous random variables map outcomes to values of uncountable sets. Of particular interest to us is the case of real-valued random variables, which are essentially functions that map outcomes to real numbers.

A *probability distribution function* (or pdf) identifies, either the probability P of each value of a random variable (when the variable is discrete), or the probability of the value falling within a particular interval (when the variable is continuous). In either case, probability distributions inherit the classification of the random variable they relate to:

- In the discrete case, we have a *discrete probability distribution*, in which case a particular value of the distribution function is denoted in upper case, as $P(\cdot)$, since it identifies directly with the probability of a specific value of the discrete random variable.
- In the continuous case, we have a *continuous probability density*, since we can only assign probabilities to intervals and not to specific values, in which case a particular value of the distribution function is denoted in lower case, as $p(\cdot)$, and the probability of the random variable being instantiated with a value within a given interval is determined indirectly by integrating the area under the density curve bounded by that interval.

Modern digital computer-controlled systems, such as robots, use analog-to-digital and digital-to-analog converters, which discretise readings providing input from sensors and control commands providing output to actuators. Hence, throughout this book most of the random variables, and consequently the respective probability distributions, will be discrete. However, we would like to maintain a certain degree of generality in this chapter on the fundamentals of probabilistic inference, and therefore, while still focussing primarily on discrete entities, we will keep addressing both realities for now.

Example 1.4. **Probability distribution of rolling a fair die**

Consider the die rolling experiment once again. Let us define a random variable C that assigns a real value to each outcome of the outcome space

[3] This is in case it is discrete; if it is continuous, read "a value within a given interval" instead.

Ω by counting the number of dots on the top face of the die; therefore, its measurable space is defined as $C \in \{1, 2, 3, 4, 5, 6\}$.

Assume that we have also established that the die is fair, and so any outcome is equally likely. The distribution on C that reflects this fact is the *uniform distribution*, in this case given by

$$P(C) = \frac{1}{6}, \quad \forall C \in \{1, 2, 3, 4, 5, 6\}.$$

Several examples of discrete and continuous probability distributions are presented in Fig. 1.3 and Fig. 1.4, respectively[4]. When describing probability distributions followed by discrete random variables supporting a small number of events, it is common to describe their distribution using *conditional probability tables (CPTs)*, which are graphically represented using discrete data histograms, such as the 2D histogram shown at the bottom of Fig. 1.3.

Formally, probability may be defined as a function from subsets of Ω to the real line $\mathcal{R}$ that satisfies the following propositions:

Axiom 1.1. Non-negativity

The probability of an event is a non-negative real number

$$P(A) \geq 0, \quad \forall A \in \Omega.$$

Axiom 1.2. Additivity

Any countable sequence of *mutually exclusive* events $A_1, A_2, \cdots$ (i.e. $A_1 \cap A_2 \cdots = \emptyset$) satisfies

$$P(A_1 \cup A_2 \cdots) = \sum_{i=1}^{\infty} P(A_i), \quad \forall A_i \in \Omega.$$

Axiom 1.3. Normalisation (assumption of unit measure)

There are no elementary events outside the outcome space; conversely, the probability that some elementary event in the entire outcome space will occur is:

$$P(\Omega) = 1.$$

These are the *Kolmogorov axioms*, named after Andrey Kolmogorov, a preeminent 20th century Soviet Russian mathematician considered the father of the modern frequentist version of Probability Theory. The three axioms bring

[4] See also our online appendix of interactive examples.

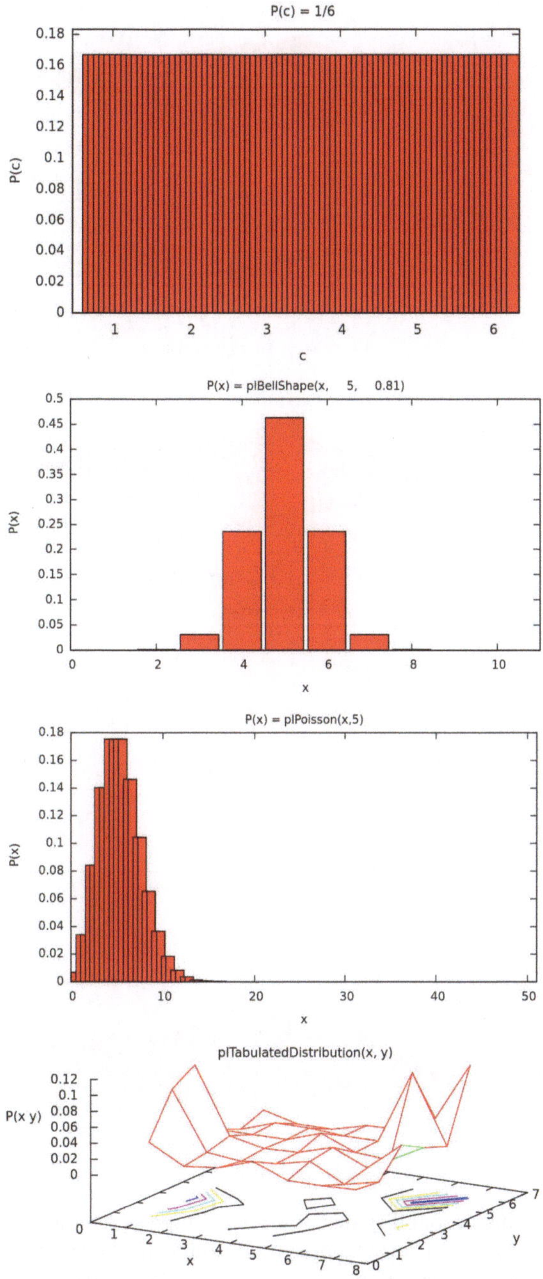

Fig. 1.3. Discrete probability distribution examples generated using the ProBT toolbox (Chapter 3). From top to bottom: 1D uniform distribution (corresponding to Example 1.4), 1D bell-shaped distribution (which follows Gaussian distribution, and is therefore sometimes called a discrete truncated Normal distribution), 1D Poisson distribution, and 2D histogram.

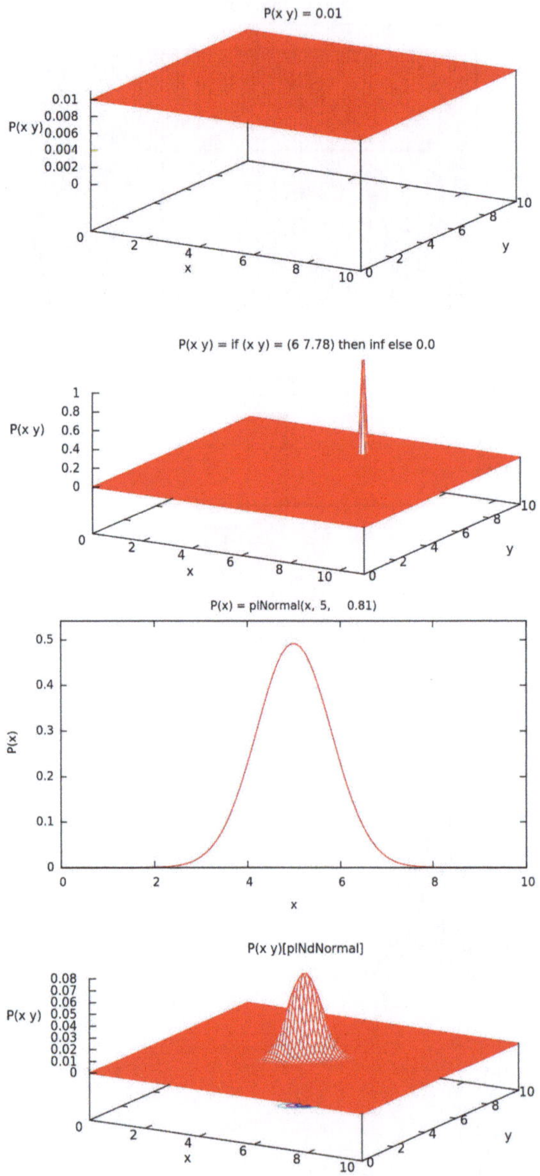

Fig. 1.4. Continuous probability distribution examples generated using the ProBT toolbox (Chapter 3). From top to bottom: 2D uniform distribution, 2D Dirac distribution, 1D Normal distribution, and 2D normal distribution.

several important consequences, the proofs of which are given in countless textbooks and beyond the scope of this text[5]:

- Probabilities always satisfy $0 \leq P(A) \leq 1$.
- A discrete random variable A always satisfies $\sum_A P(A) = 1$.
- Analogously, a continuous random variable A always satisfies $\int_A p(A)dA = 1$.
- The probability of two arbitrary *combined events* (i.e. in general non-mutually exclusive) A and B, denoted as $P(A \cup B)$, is given by the *generalised probability addition rule*

$$P(A \cup B) = P(A) + P(B) - P(A \cap B), \tag{1.3}$$

  where $P(A \cap B)$ represents the probability that both occur (i.e. their *joint probability*).

1.2.2 Marginal Distributions, Independence and Conditional Probability

The *marginal distribution* of a subset of a collection of random variables is obtained by *marginalising* (i.e. summing or integrating, in the discrete or continuous case, respectively) over the distribution of the variables being discarded — the discarded variables are said to have been *marginalised out*. The term *marginal variable*[6] is used to refer to those variables in the subset of variables being retained.

Thus, in the discrete random variable case,

$$P(S) = P(S \cap \Phi_1) + \cdots + P(S \cap \Phi_N), \tag{1.4}$$

where S and $\Phi_1 \cdots \Phi_N$ are the marginal variable and the discarded variables, respectively, representing the collection Ω, for which, according to Axiom 1.5, $P(\Omega) = 1$.

The *conditional probability* of A given that B occurred, denoted as $P(A \mid B)$, is the joint probability of A and B, divided by the marginal probability of B:

$$P(A \mid B) = \frac{P(A \cap B)}{P(B)}. \tag{1.5}$$

The conditional probability of A knowing B occurred, if these events are *independent* from each other, is obviously equivalent to the marginal

[5] "Mercifully!", we admit was your most probable thought...

[6] As a historical curiosity, these terms are dubbed "marginal" because they used to be found by summing values in a table along rows or columns, and writing the sum in the margins of the table.

Table 1.1. Commonly used statistics

Name	Notation	Expectation
mean (expected value)	μ_X	$E[X]$
variance (squared standard deviation)	σ_X^2	$E[(X - E[X])^2] = E[X^2] - E[X]^2$
covariance	$Cov[X, Y]$	$E[(X - E[X])(Y - E[Y])] = E[XY] - E[X]E[Y]$
correlation	$Cor[X, Y]$	$\frac{Cov[X,Y]}{E[X]E[Y]}$

probability of A, $P(A \mid B) = P(A)$. Substituting into (1.5) and solving for $P(A \cap B)$, the joint probability of two independent events is obtained

$$P(A \cap B) = P(A)P(B). \tag{1.6}$$

1.2.3 Statistics and Expectation

We have shown that random variables *follow* probability distributions. The set of quantified characteristics of these variables taken from observing their values repeatedly are called *statistical measures*. They frequently serve as *estimators* for the *parameters* of the underlying probability distribution, which are not computable since often the population is generally much too large to examine – more on estimation in the following subsection. Statisticians commonly try to describe the observations using measures of location, or central tendency, such as the mean, a measure of statistical dispersion, such as the standard deviation, a measure of the shape of the distribution, such as skewness or kurtosis, and if more than one variable is considered, a measure of statistical dependence such as a correlation coefficient.

Statistical moments are computed using an operation called the *expectation* of a function $f(X)$ of a random variable X following a distribution $P(X)$

$$E_{P(X)}\left[f(X)\right] = \sum_{i=1}^{N} P(x_i)f(x_i). \tag{1.7}$$

The shorthand notations $E_X[f(X)]$ or even $E[f(X)]$ are usually used when the underlying distribution is unequivocal.

The most common statistical measures obtained by taking the expectation are given in Table 1.1.

1.2.4 Frequentist Inference

According to the frequentist view, as we have seen before, inference should always be interpreted and evaluated in terms of a result yielded from hypothetical repetitions under the same conditions – "what would happen if we did this many times?" Unfortunately, perception, as most of cognition, aims to infer properties of the observed world from incomplete data; perceptual beings, to be able to cope with the dynamics of the observed ever-changing world, need to start performing inference right from the get-go, albeit refining their knowledge over the surrounding environment by incrementally updating their internal representations. This means that inference, in this context, must be performed even if a specific perceptual scenario is encountered only **once**. So why mention frequentist inference at all?

There are two basic types of frequentist inference:

1. **Estimation** – inferring a plausible range of values for unknown population parameters.
2. **Testing** – deciding whether hypotheses concerning values of unknown population parameters comply with sample data.

Estimation, specifically, has particular relevance for a very important aspect of probabilistic approaches to artificial perception: learning of unknown distribution parameters from training data (the rough equivalent of the learning by experience process used by humans during development – see Chapter 6). It also plays a major role in decision making, since one could say that it involves inferring a concrete value, given a distribution, that best fulfils a specific goal, and also since it is conceptually the companion process of learning (see Chapter 5).

Point estimation is perhaps the simplest, yet simultaneously the most ubiquitous, particular case of statistical inference. It uses a single number to estimate the unknown parameter (i.e. the *point estimate*). Assuming a generic parameter θ, if $\hat{\theta}$ is its point estimate, then the estimation error $e = \hat{\theta} - \theta$ is a random variable, which, for the estimate to be as accurate as possible, should be close to zero.

Example 1.5. **Point estimation – least-squares method**

The least-squares point estimate of population mean μ is the number $\hat{\mu}$ for which the sum of squared errors $(\hat{\mu} - x_i)^2$ is at a minimum, where x_i represent the M values taken from the experimental data set X, extracted as a sample from the population.

Using the derivative of this sum with respect to $\hat{\mu}$ and equalling it to zero,

$$\frac{\partial}{\partial \hat{\mu}} \sum_{i=1}^{M} (\hat{\mu} - x_i)^2 = 0 \Rightarrow \hat{\mu} = \frac{\sum_{i=1}^{M} x_i}{M} = \bar{X}$$

This means that the average of the sample set $\bar{X}$ represents a possible estimate for the population mean μ.

Generically, if $X = \{x_1, \ldots, x_m\}$ is a set of independent observations from a probability distribution $P(X \mid \theta)$, where θ is the parameter we wish to estimate, then the probability that the full set of observations occurs would be given by

$$P(X \mid \theta) = \prod_{i=1}^{m} P(x_i \mid \theta). \tag{1.8}$$

Consequently, if one wishes to find a point estimate for θ, a reasonable criterion would be to establish that the estimate should maximise $P(X \mid \theta)$; in simple words, to compute the value $\hat{\theta}$ for which $P(X \mid \theta)$ is higher than for any other value of θ. For this purpose, it is useful to define a *likelihood function*, that measures the relative likelihood that different θ have resulted in the observed X as

$$L(\theta) \propto \prod_{i=1}^{m} P(x_i \mid \theta). \tag{1.9}$$

If we want to estimate θ, we want to find a particular $\hat{\theta}$ which maximises L. A broadly used inference strategy for point estimation, known as Maximum Likelihood Estimation (MLE), can be described as the process of choosing the value for the distribution parameter that *maximises the likelihood of the observations*,

$$\hat{\theta} = \arg\max_{\theta} L(\theta) = \arg\max_{\theta} \prod_{i=1}^{m} P(x_i \mid \theta), \tag{1.10}$$

as the desired point estimate for θ.

As in Example 1.5, and for many estimation problems, this equation can be solved by differentiating $L(\theta)$ with respect to θ, equalling it to zero, then solving for θ (checking second order derivatives to ensure a global maximum is obtained). This corresponds to computing

$$\frac{\partial}{\partial \theta} P(X \mid \theta) = \frac{\partial}{\partial \theta} \prod_{i=1}^{m} P(x_i \mid \theta) = 0. \tag{1.11}$$

Since maximising the likelihood function is equivalent to maximising its logarithm, $l(\theta) = \ln L(\theta)$, Equation (1.10) becomes

$$\begin{aligned}\hat{\theta} &= \arg\max_{\theta} l(\theta) \\ &= \arg\max_{\theta} \ln\left(\prod_{i=1}^{m} P(x_i \mid \theta)\right) \\ &= \arg\max_{\theta} \sum_{i=1}^{m} \ln\left(P(x_i \mid \theta)\right),\end{aligned} \tag{1.12}$$

which is clearly much less expensive computationally.

Example 1.6. **Point estimation – Maximum Likelihood Estimation (MLE) method for a binomial distribution**

Imagine that we are trying to characterise the performance of a faulty infrared sensor in detecting the presence of obstacles, based on a set of sensor readings, $x_i = \{x_1, \cdots, x_n\}$ taken in n *independent repetitions* of the *same obstacle detection experiment* in which *we know an obstacle was present.*

Denoting the probability of the occurrence of a successful detection D as θ (and, consequently, the probability of failure F as $1-\theta$), our task would be to estimate the value of the parameter θ. To do this, we need to define the parameter space, which in fact also a hypothesis space (more on this concept in Chapter 6), and also an *objective function* to help us decide how accurately the different hypotheses in the parameter space characterise the sensor readings we collected. In this case, the parameter space is given by the set of all parameters $\theta \in [0,1]$, and the probability distribution followed by X is said to be *binomial.*

Using the MLE criterion as our objective function, if the detection success data are likely given a parameter value, that parameter value is a good predictor for the data, and thus represents a good point estimate.

Imagine we observe the sequence of detection outcomes D, F, F, D, D. If we knew θ, we could assign a probability to observing this particular sequence. The probability of the first detection attempt is given by $P([x_1 = D]) = \theta$. Given our independent trials assumption, the probability of the second attempt is given by $P([x_2 = F] \mid [x_1 = D]) = P([x_2 = F]) = 1 - \theta$, and so on. Thus, the probability of the full sequence is

$$P(X = \{D \wedge F \wedge F \wedge D \wedge D\} \mid \theta) = \theta^3(1-\theta)^2.$$

Note that, again due to the independence assumption, the actual ordering of the outcomes is irrelevant.

What would be the likelihood function for the general case of this set of experiments, and how could we perform MLE using it? Assume that our detection success dataset Δ contains exactly $\Delta\#(\alpha)$ success events, denoted as $\alpha = D$. In this case, the likelihood function is given by

$$L(\theta) \propto \theta^{\Delta\#(\alpha)} (1-\theta)^{n-\Delta\#(\alpha)}.$$

Resorting to the log-likelihood,

$$l(\theta) \propto \Delta\#(\alpha) \ln \theta + (n - \Delta\#(\alpha)) \ln (1-\theta).$$

Differentiating the log-likelihood, solving with respect to θ and setting the derivatives to zero, we obtain the MLE estimate

$$\hat{\theta} = \frac{\Delta\#(\alpha)}{n}.$$

Example 1.7. **Point estimation – Maximum Likelihood Estimation (MLE) method for a normal distribution**

Imagine that we are trying to characterise the performance of a sonar sensor in measuring a specific distance, based on a set of sensor readings. In practice, we have a finite number of measured values $x_i = \{x_1, \cdots, x_n\}$, and our problem is to arrive at the "best estimates" of μ and σ based on these readings.

If we consider n independent readings taken by the sonar — the ith of which being denoted as x_i — and if we assume that the sonar readings will follow a normal distribution $\mathcal{N}_{\mu,\sigma}(x)$ centred around the real distance value, the likelihood of our sample would be

$$L(\theta_{\mu,\sigma}) \propto \prod_{i=1}^{n} P(x_i \mid \theta_{\mu,\sigma}) = \prod_{i=1}^{n} \frac{1}{\sigma\sqrt{2\pi}} e^{-\frac{(x_i-\mu)^2}{2\sigma^2}}.$$

Since $\sqrt{2\pi}$ is constant, it can be factored out, resulting in

$$L(\theta_{\mu,\sigma}) \propto \frac{1}{\sigma^n} e^{-\sum_{i=1}^{n} \frac{(x_i-\mu)^2}{2\sigma^2}}.$$

Resorting to the log-likelihood,

$$l(\theta_{\mu,\sigma}) \propto -\sum_{i=1}^{n} \frac{(x_i-\mu)^2}{2\sigma^2} - n \ln \sigma.$$

Differentiating with respect to μ and setting the derivatives to zero,

$$\frac{\partial l(\theta_{\mu,\sigma})}{\partial \mu} = \sum_{i=1}^{n} (x_i - \mu) = 0.$$

Solving for μ we find that

$$\hat{\mu} = \frac{\sum_i x_i}{n} = \bar{X}.$$

To find the best MLE estimate for σ

$$\frac{\partial l(\theta_{\mu,\sigma})}{\partial \sigma} = \frac{\sum_{i=1}^{n}(x_i - \mu)^2}{\sigma^3} - \frac{n}{\sigma} = 0.$$

The true value for μ is unknown, and therefore, in practice, it is replaced by its best MLE estimate $\bar{X}$. Consequently, the maximum happens at

$$\hat{\sigma} = \sqrt{\frac{1}{n}\sum_{i=1}^{n}(x_i - \bar{X})^2}.$$

Be aware that, depending on the method used, different estimates may be derived. In statistics, generally the *unbiased* version (the explanation of which is beyond the scope of this book) of $\hat{\sigma}$

$$\hat{\sigma} = \sqrt{\frac{1}{n-1}\sum_{i=1}^{N}(x_i - \bar{X})^2},$$

is used. Note, however, that, in this particular case, for a large number of samples $n >> 1$, both estimates become approximately equal.

1.3 Bayesian Inference and Modelling

1.3.1 Plausible Reasoning and Propositions

Jaynes [4], in his seminal posthumously published book on probability theory, used the parable of a "thinking robot", an imaginary being's brain, to present a novel approach to logic called *plausible reasoning*. He started by stating that this thinking robot would reason about *propositions*, which, according to the author, are general statements (which can be qualitative or quantitative in nature) that must have an *unambiguous meaning* and be of the "simple, definite logical type that must be either true or false", alluding to the two-valued logic introduced by Aristotles. These propositions are denoted as italicised capital letters, $\{A, B, C, \text{etc.}\}$, unless otherwise stated.

Then he produced a further statement which is of utmost importance to us: "we do not require that the truth or falsity of such an 'Aristotelian proposition' be ascertainable by any feasible investigation; indeed our inability to do this is usually just the reason why we need the robot's help" [4]. So, as an alternative to symbolic logic (commonly known as *Boolean algebra*, acknowledging George Boole's contribution to this subject), which assumes absolute knowledge over the truth or falsity of propositions, he proposed that the artificial brain he was hypothetically designing would only have the ability of assigning a degree of *plausibility*, based on the evidence gathered about these propositions, reviewing these assignments as new evidence is acquired and accumulated.

With this intent in mind, Jaynes proposed a hypothetical design of such a brain by deriving rules from three basic desiderata:

(I) *Degrees of plausibility are represented by real numbers.*
(II) *Plausibility should have a qualitative correspondence with common sense.*
(III) *If the plausibility of a proposition can be derived in different ways, all results must be consistent.*

We can now define probability in the context of plausible reasoning according to Jaynes, denoted as $P(A \mid B)$, as the assignment of a degree of plausibility to some proposition A depending on some other proposition B, whose veracity is assumed to be known [4]. Note that, according to this definition, probability *always* implies the assignment of a *conditional plausibility*. However, the acknowledgement of the dependence of A in B is sometimes made implicit (e.g. B might be a latent variable), and in these cases the notation for probability is simplified to $P(A)$.

Parallely, Pearl [5] showed that, for plausible reasoning to be a realistic alternative to symbolic logic, which allows computing the impact of each new fact in decomposable *stages*, it too must emulate this ability — this can only be justified when taking uncertainty into account by making some restrictive assumptions of independence. Therefore, both the notion of dependence and the notion of independence represent the foundations of plausible reasoning.

Returning to the desiderata presented above, these are, in fact, a set of postulates which were declared by physics professor Richard Threlkeld Cox in his well-known theorem. From Cox's postulates one can obtain Cox's axioms, as derived by Jaynes [4]:

Considering propositions A, B and C:

Axiom 1.4. Certainty

Certain truth is represented by $P(A \mid B) = 1$, and certain falsehood by $P(A \mid B) = 0$.

Axiom 1.5. Negation/Normalisation

Denoting the negation of proposition A as $\bar{A}$,

$$P(A \mid B) + P(\bar{A} \mid B) = 1.$$

Axiom 1.6. Conjunction

Denoting the conjunction of two propositions as $A \wedge B$ (conceptually equivalent to the conjunction of events presented earlier, although denoted differently, since events are sets and propositions are logical entities; sometimes, in shorthand, the $\wedge$ symbol is ommited),

$$P(A \wedge B \mid C) = P(A \mid C)P(B \mid A \wedge C) = P(B \mid C)P(A \mid B \wedge C).$$

From these axioms, the Kolmogorov axioms presented in section 1.2 can be derived; however, plausible reasoning does not interpret propositions in terms of sets, but probability distributions as carriers of incomplete information [4]. As a result, the Kolmogorov system's scope of application does not include the so-called *ill-posed and generalised inverse problems*, which form the basis of most of the difficulties encountered in mathematically modelling perception, while plausible reasoning *does* [4].

Propositions can be relatively abstract or even qualitative statements (which, nonetheless, should be unambiguously either true or false); however, it is desirable in our case that propositions relate to concrete facts or, even better, to concrete *values*. So, if a specific proposition was to be related to a random variable, it should reflect the veracity of that variable being instantiated with a specific value. Consequently, if we would wish to represent the degree of plausibility of a random variable A being instantiated with value a, fact notated as $[A = a]$, the probability of that value occurring given a context reflected by proposition B is written as $P([A = a] \mid B)$. Discarding the influence of B (again, B might be a latent variable, or it might have already been instantiated through $[B = b]$, and may consequently be deemed as irrelevant), the shorthand representation for the degree of plausibility assigned is denoted in this case as $P(A = a)$. Consequently, by abuse of notation, the probability of an unspecified value occurring for A, in other words, the degree of plausibility assigned to unspecified proposition A stated regarding knowledge on random variable A and its respective measurable space, is denoted as $P(A)$. This implies that random variable A follows the probability distribution denoted by $P(A)$.

So, in summary:

- To simplify reading and introduce homogeneity into the notation, single probability values, discrete probability distributions and families of discrete probability distributions are all generically formally denoted as conditional probabilities, $P(A \mid B)$. They are distinguished from one another by the context of their arguments, described next.
- Using this notation, $P(A \mid B)$ is a *family of probability distributions*, one distribution for each possible value of B; $P(A \mid [B = b])$ is one such distribution; and $P([A = a] \mid [B = b])$ is a single probability value corresponding to a specific proposition.
- In exceptional cases, there are only dependences on hidden or instantiated variables, in which case the notation may be reduced to $P(A)$.

This important practical correspondence between propositions and variables was introduced by Bessière, Laugier, and Siegwart [1]. A discrete random variable X, according to these authors, can be alternatively defined, in the context of plausible reasoning, as a set of mutually exclusive ($\forall i, j$ with $i \neq j$, $[X = x_i] \wedge [X = x_j]$ is false) and exhaustive (at least one of the instantiations $[X = x_i]$ will be true) logical propositions, with $\langle X \rangle$ denoting the cardinality of the set X (i.e. the number of propositions $[X = x_i]$ of that set, or, equivalently, the size of the measurable space of X).

The conjunction of two variables X and Y, denoted as $X \wedge Y$, is defined as the set of $\langle X \rangle \times \langle Y \rangle$ propositions $[X = x_i] \wedge [Y = y_j]$. $X \wedge Y$ is a set of mutually exclusive and exhaustive propositions, and, as such, is itself also a new discrete random variable. This rationale is recursive: any conjunction of n variables can be considered at any time as a single discrete random variable.

Contrastingly, the disjunction of two variables, defined as the set of propositions $[X = x_i] \vee [Y = y_j]$, is *not* a random variable, since the propositions are not mutually exclusive.

The way of defining probability presented above is certainly controversial; in the frequentist view, for example, as described in section 1.2, probabilities correspond to frequency counts in experiments performed repeatedly, as many times as possible; consequently, according to partisans of this view, the notion of "plausibility" as an *objective* representation of real, physical phenomena might be construed as preposterous... In response, Pearl [5] and several other authors proposed an alternative definition of probability: in the context of plausible reasoning, probabilities would represent the degree of *subjective belief* of the cognitive system over propositions; the mathematical background would, otherwise, be basically the same as what proposes Jaynes and similarly thinking authors.

Consequently, whether one would opt for the definition of probability as the degree of plausibility assigned by the thinking brain to a proposition — assuming that, although taken from a subjective point of view, the knowledge constructed in such a way would be unbiased by personal creeds, if any, of the cognitive system (or its programmer), and thus construed as objective — or as the degree of subjective belief of that cognitive system over the same proposition, is essentially irrelevant in terms of implementation. Therefore, for lack of a better term, we will refer to both as *Bayesian approaches*, and we will deal with their common grounds in the text that follows.

1.3.2 Bayes' Theorem and Bayesian Inference

Let us now assume that we would like to apply plausible reasoning so as to use proposition B, of which the veracity is assumed to be known, to infer the plausibility concerning proposition A, by relating it conditionally to B; put in formal and concise terms, we would like to compute $P(A \mid B)$. Let us also define the *context* of a conditional probability as the conjunction of variables behind the conditioning bar $|$.

Any given model is incomplete: a set of factors within its context are always unaccounted for that influence the observed phenomenon, either because some of them are difficult or even impossible to know or predict, or because they are just too complex to model efficiently. The consequence of these factors, represented by *latent variables* and denoted collectively as π, is that the model **cannot** fully describe the underlying phenomenon.

A robot designer has an abstract conception of the environment that surrounds the robot and its sensors; they might be described in geometrical terms, because the geometry of the entities in the world can be specified, in analytical terms because the laws of physics that govern this world are assumed to be known, and in symbolic terms, because both the objects and their characteristics can be named. So, in fact, the designer imposes on the robot his or her own abstract conception of the physical world.

Unfortunately, this results in an "irreducible incompleteness of the model" [1]: since the model does not completely account for the influence of the context on the robot's sensors, in an odd reversal of causality, sensor data is usually described as "noisy" (see Fig. 1.5), as if due to an imperfect physical world, while the model is assumed as correct.

In fact, just the opposite is true: there is no noise in the physical world, but there is a substantial amount of simplification and inherent ignorance underlying perceptual models. Uncertainty can then be recognised for what it is: a consequence of the lack of knowledge, a direct sub-product of incompleteness [1][7]. So, how can plausible reasoning help under these circumstances?

Reverend Thomas Bayes (1702–1761) lent his name to the concrete inference approaches to plausible reasoning through the following theorem:

[7] See section 1.4.1 for the formal mathematical definition of uncertainty as the consequence of the lack knowledge — relating it to the amount of *information* gathered over a phenomenon.

Fig. 1.5. Latent variables (denoted collectively as π) in the context of perception. Many factors, unaccounted for in most cases, affect perceptual models: examples of internal factors would be accuracy and precision ratings, discretisation due to analogue-to-digital transformations, approximation truncations, and round-off-effects from numeric representation limitations (i.e., finite number of digits used by digital memories and processing units); while external factors would be, for example, ambiguity due to physical constraints (e.g. the mapping of 3D objects into 2D images, or the "aperture problem" in local motion detection) or illumination conditions, etc. What different latent variables, commonly construed as "sensor noise", can you identify in the perceptual robot's predicament represented above?

Theorem 1.1. ***Bayes' theorem (also known as Bayes' law or rule)***

$$P(A \mid B) = \frac{P(A)P(B \mid A)}{P(B)}.$$

Proof. Explicitly acknowledging the effect of latent variables on the context and applying Cox's conjunction axiom (axiom 1.6 on page 21), we have that

$$P(A \wedge B \mid \pi) = P(A \mid \pi)P(B \mid A \wedge \pi) = P(B \mid \pi)P(A \mid B \wedge \pi).$$

It immediately follows that

$$P(A \mid B \wedge \pi) = \frac{P(A \mid \pi)P(B \mid A \wedge \pi)}{P(B \mid \pi)},$$

which, using shorthand notation by making the effect of latent variables implicit, proves Bayes' theorem as stated above.

□

Bayes' theorem is accompanied by several related and very significant definitions and notions. Examining Bayes' rule more carefully, when entering the inference process to determine $P(A \mid B)$, we are faced with what is called *prior* or *a priori* on A, denoted as $P(A)$, the *likelihood*, denoted as $P(B \mid A) = L(A)$, which stands for the measured plausibility of the known proposition B given A, and the so-called *evidence* collected concerning the known proposition B, denoted as $P(B)$. Therefore, since $P(A \mid B)$ is the *result* of inference given B, it is known as the *posterior* or *a posteriori* on A.

Note that since the posterior is proportional to both the likelihood[8] and the prior and the evidence serves as a normalisation factor, it is common to write the posterior as

$$P(A \mid B) \propto P(A)P(B \mid A). \tag{1.13}$$

This fact is often used to simplify computations, in applications where the actual value of a specific $P(A \mid [B = b])$ is not important, only its relative weight in comparison to posteriors for other values for B.

Let us now consider a simple example of applying Bayes' theorem in plausible reasoning before advancing into further considerations over the full range of possibilities made available by the Bayesian approach.

[8] Cf. with the definition of the likelihood function in section 1.2.4.

Example 1.8. **Simple example of Bayesian inference for plausible reasoning**

Consider that we have built a robot that reasons over diagnostic tests T performed on industrial equipment, so as to decide if it should be sent for repairing. These tests, for each piece of equipment, yield either 1, if the equipment is faulty, or 0, if no fault has been detected. However, these tests only assess the apparatus' performance indirectly, by observing the machine's behaviour during its operation, so a degree of uncertainty in these assessments is assumed. The proposition to be assessed is, therefore, $F =$ "equipment is out of order" knowing the result for T; by abuse of notation, let us consider a binary random variable F, which is 1 when this proposition is known to be true and 0 when it is false.

Now, imagine that this robot knows that only 0.1% of this type of equipment is faulty at a given time, i.e. $P(F) = .001$. The robot also knows that the test's ratings (i.e. the probability of the test yielding a specific result knowing if the equipment is faulty or in perfect working conditions) are given by the following probability table for $P(T \mid F)$:

	$[F=0]$	$[F=1]$
$[T=0]$	.98	.05
$[T=1]$	.02	.95
$\sum_T P(T \mid F)$	1	1

Will the robot be capable of performing Bayesian inference? What would it decide if the test yielded positive?

Bayes' rule states that, in this situation,

$$P([F=1] \mid [T=1]) = \frac{P([F=1])P([T=1] \mid [F=1])}{P([T=1])} = \frac{.001 \times .95}{P([T=1])}.$$

Apparently, the robot has no means of assessing $P(T)$, the evidence on T. However, by consecutively applying marginalisation over all possible states of F – equation (1.4) – and the definition of conditional probability – equation (1.5) – the robot is capable of arriving at the conclusion that

$$\begin{aligned} &P([F=1] \mid [T=1]) = \\ &\quad = \frac{P([F=1])P([T=1] \mid [F=1])}{P([T=1] \wedge [F=0]) + P([T=1] \wedge [F=1])} \\ &\quad = \frac{P([F=1])P([T=1] \mid [F=1])}{P([F=0])P([T=1] \mid [F=0]) + P([F=1])P([T=1] \mid [F=1])}, \end{aligned}$$

and thus

$$
\begin{aligned}
P([F=1] \mid [T=1]) &= \\
&= \frac{.001 \times .95}{.999 \times .02 + .001 \times .95} \\
&\approx .019\,,
\end{aligned}
$$

or, in other words, that the probability of the equipment being out of order given a positive test result is 1.9%.

Note that the robot's prior knowledge had a huge influence on the posterior, even in the presence of such a strong likelihood.

The likelihood $P(B \mid A)$ and the prior $P(A)$ are usually known, measured or fixed in advance; the summarised evidence $P(B)$, however, is usually not immediately available and, as in Example 1.8, in that case one has to resort to marginalisation over all possible states of the unknown variable.

As can be understood from the previous example, Bayes theorem provides an interpretation of sensor data in respect to the hypothesis regarding the sensorial scenario coded into a model (we will come back to this with further detail in Chapter 2):

$$
P(\text{hypothesis} \mid \text{data}) = \frac{P(\text{data} \mid \text{hypothesis})P(\text{hypothesis})}{P(\text{data})} \tag{1.14}
$$

Bayes' theorem in this case therefore yields the degree of plausibility or belief of a hypothesis based on prior knowledge on that scenario considering the hypothesis, $P(\text{hypothesis})$, initially embedded into the robot's cognitive system by the designer, and on the designer's knowledge on the probability of a sensor reading assuming the veracity of the hypothesis, $P(\text{data} \mid \text{hypothesis})$. It is **essential** to note that preliminary knowledge comes in two flavours, *encoded simultaneously in the prior and in the likelihood*, both of which can be hard-coded or learned through experience (more on this in Chapter 6). This degree of plausibility, at the moment it is inferred, is denoted as $P(\text{hypothesis} \mid \text{data})$.

1.3.3 Markov Processes and the Markov Property

The preliminary knowledge used by the robot is moulded by the data being observed by the robot and then reused in later inference steps — this means that Bayesian inference was designed to be performed *recursively*. However, this recursion depends on how the probabilistic process is to be modelled in terms of its temporal evolution, namely on how the conditional probability distribution of its future states, conditional on both past and present states, *effectively* depend on the present and previous states.

Modelling a probabilistic process in such a way that the the conditional probability distribution of its state taken at any time instant depends on all its predecessors leads to an intractable situation as time progresses, requiring more and more memory and more and more computations.

One way to deal with this problem is to model the process according to the assumption that it approximately follows the *Markov property*. A probabilistic process has the Markov property if the conditional probability distributions of its future states depend *only* on its present state, and not on past states.

In the case that the process state ${}^{t}X$ can take on a discrete set of values at a particular time instant t when inference is performed, then the Markov property holds, and the process is called a *Markov process*, if the following condition holds

$$P({}^{t+1}X \mid {}^{t}X \wedge {}^{t-1}X \wedge {}^{t-2}X \wedge \ldots) = P({}^{t+1}X \mid {}^{t}X). \tag{1.15}$$

with t denoting the current inference step. In this case, t can be considered as a discrete representation of time; however, this condition may be easily rewritten so as to generalise to the case where t is a continuous variable.

It follows that, if the process has the Markov property, the recursive application of Bayes rule becomes computationally trivial, and the update can be simply carried out using $P({}^{t}\text{hypothesis}) \equiv P({}^{t-1}\text{hypothesis} \mid {}^{t-1}\text{data})$.

Let us expand the concepts of current and future states, namely by introducing an intermediate variable Y representing a time-interval of states X; if Y has the Markov property, the apparently non-Markovian process X is, in fact, a *second-order Markov process* , and Y is considered a *first-order Markov process*, where the order of the process represents the number of levels of temporal dependence through the Markov property. This idea can be generalised in order to define *Markov processes of order n*.

Example 1.9. **Simple example of sensor modelling using Bayesian inference for plausible reasoning**

Consider that we have built a robot that reasons over *independent* sonar readings s^t at each time-step t in order to determine the distance X to an obstacle in front of the robot. Assume that all random variables in the following are discrete and integer, with minimum and maximum equalling 0 and 10,000, respectively, and that they are referred to in millimetres.

Now imagine that sensor readings yielded by the sonar mounted on the robot follow a probability distribution as described in Example 1.7, and that the distribution parameters are estimated as described in that example, and found to be $\mu = X$ (i.e. the distribution is centred on the real value) and $\sigma = 5$ (i.e. the average error incurred by the sensor is 5 mm).

If the robot senses, using the sonar, two different values given by $s^t \in \{1000, 1500\}$ at two respective consecutive instants $t \in \{0, 1\}$ for a given obstacle, which is assumed to be stationary but for which no other information is known in advance, what will be the distance X inferred by the robot's perceptual system?

Let us begin by systematically organising and stating what we know about the problem.

1. We know that the obstacle is always the same object and that it remains stationary, so the hypothesis being tested, $X^0 = X^1 = \ldots = X$, is always the same.
2. We know that the final estimate, resulting from inference, can be obtained by recursively applying Bayes' rule for each time-step t, and by consequently taking the value for X corresponding to the highest probability given by the posterior.
3. We know nothing about the distance to the obstacle at the beginning of the robot's reasoning ($t = 0$), which means that our first prior is given by a uniform distribution in X, i.e. $P(X) = \mathcal{U}(X)$.
4. We know that the likelihood for sonar readings is given by $P(s^t \mid X) = \mathcal{N}(\mu = X, \sigma = 5)$, which is a family of normal distributions centred on each value of X. Therefore, if one considers a specific value s^t, the corresponding likelihood function $L(X)$ is given by the result of the convolution of a discrete-time unit impulse displaced to $X = s^t$ (i.e. a discrete displaced Delta function) with the normal distribution described above, resulting (due to the shift-invariance property of convolutions) in a normal distribution with $\mu = s^t$ and $\sigma = 5$.

Using Bayes' rule in this fashion yields, for the first time-step $t = 0$,

$$P(X \mid s^0) = \frac{P(X)P(s^0 \mid X)}{P(s^0)}.$$

Given that $P(X)$ is uniform,

$$P(X \mid s^0) \propto P(s^0 \mid X).$$

To obtain an estimate for X, a sensible strategy (but not the only one – refer to Chapter 5) would be to compute $\hat{X}$ corresponding to the maximum for the posterior. In this case, this would result in computing the maximum of the likelihood function $L(X)$, which, since likelihood is a normal distribution, would fall on the mean. Therefore, $\hat{X} = s^0 = 1000\,\text{mm}$, with an average estimation error in distance of $\sigma = 5\,\text{mm}$.

For the following time-step, $t = 1$, using the previous posterior as the new prior and knowing that sensor readings are independent from one another, the Bayesian update process results in

$$P(X \mid s^1) = \frac{P(X)P(s^1 \mid X)}{P(s^1)} = \frac{P(X \mid s^0)P(s^1 \mid X)}{P(s^1)}.$$

Substituting with the result of Bayesian inference in the previous time-step,

$$P(X \mid s^1) \propto P(s^0 \mid X)P(s^1 \mid X).$$

The product of normal distributions is known to also be a normal distribution. In the case of a unidimensional normal distribution, the mean of the final distribution is given by the sum of the means of each distribution weighted by the respective standard deviations (i.e. weighted average of the means, with standard deviations as weights),

$$\mu_{final} = \frac{\mu_0 \sigma_0 + \mu_1 \sigma_1}{\sigma_0 + \sigma_1},$$

while the corresponding standard deviation is given by

$$\sigma_{final} = \sqrt{\frac{\sigma_0^2 \sigma_1^2}{\sigma_0^2 + \sigma_1^2}}.$$

Given that the maximum of a normal distribution is found at the mean, and that the likelihood standard deviations are all equal, the robot's estimate for the distance to the obstacle is given by substituting each mean by the respective sensor reading value, resulting in

$$\hat{X} = \mu_{final} = \frac{1000 + 1500}{2} = 1250\,\text{mm},$$

which corresponds to an average estimation error in distance of

$$e_X = \sigma_{final} = \sqrt{\frac{5^2 \times 5^2}{5^2 + 5^2}} = 3.5355\,\text{mm}.$$

The robot has thus refined its estimate for the distance to the obstacle from the initial time-step $t = 0$ to the next time-step $t = 1$. On the other hand, σ_{final} is lower than the standard deviation of the posterior computed for $t = 0$. This means that the uncertainty of the robot's perceptual system has lowered from the first time-step to the next, implying that the robot can be more confident on its current estimate for X.

This process can now be repeated, as new readings arrive from the sonar sensor.

1.3.4 Bayesian Modelling

We can finally answer the question of incompleteness and achieve full circle concerning Fig. 1.1 on page 5: we have shown that plausible reasoning and the Bayesian approach deal with incompleteness by *summarising the influence of latent variables on the context of a model through the uncertainty encoded in the distributions involved in the inference.* So, although these latent factors are diluted and made implicit, they are nevertheless present in the model.

We are now also in the position of stating that more complicated plausible reasoning is divided, in general, into three stages:

Forward modelling: Firstly, all parameters are expressed as random variables, all independence assumptions are stated, and parametric forms of conditional probability distributions are defined by establishing/designing the prior(s) and likelihood(s).

Establishing the generative model: Secondly, the degree of plausibility of all parameters is summarised by defining the joint probability distribution, therefore establishing the so-called *generative model* (more on the significance of this in Chapter 2).

Bayesian inference: *Bayesian inference* is then performed on the generative model by following the systematic application of the basic rules presented above (therefore, implicitly, Bayes' rule): the conjunction rule (axiom 1.6), the normalisation rule (axiom 1.5) and the marginalisation rule, given by equation (1.4) (itself derivable from the first two rules).

A set of sophisticated and systematic approaches for applying these two stages is presented on Chapter 3.

1.4 Information and Sensory Processing

1.4.1 Information and Entropy

Perception can be defined loosely as the act of processing sensorial information. But what is the quantitative definition of "information"?

Consider the event of observing a particular value x for a random variable X. How informative is this information given the we know the probability of the occurrence of that event as being $P(X = x)$? The added value of an informative event should be its capacity to question the current state of knowledge of a cognitive system, or, in other words, its ability to elicit "surprise". How much "surprise" is generated from an event $[X = x]$ — how can we establish the relation between that "surprise" and the probability of the event occurring?

Shannon quantified this "surprise" as the logarithm of the inverse of the probability of that event

$$H(X = x) = \log \frac{1}{P(X = x)} = -\log P(X = x). \tag{1.16}$$

This means that information would be zero when the event is completely predictable, $P(X = x) = 1$, and increases as $P(X = x)$ decreases. The justification for applying logarithm is analogous to what was used above for log-likelihood: the information yielded by a joint distribution of independent random variables is given by the sum of each event. In many cases it is convenient to use the binary logarithm $\log_2$, and in this case the unit of information is referred to as a *bit* (short for "binary digit").

As the act of observing values of X is repeated, x is *expected* to follow $P(X)$, so the average information by repeating this process, the *entropy* of X, is given by

$$H(X) = E[-\log P(X)] = \sum_{X} -P(X) \log P(X). \tag{1.17}$$

Imagine that we have a variable X for which we know with absolute certainty the value it will be assigned within its measurable space even before the assignment is made. In fact, this limit-case effectively removes from that variable the essence of being random – it is now a deterministic variable. In any case, it is still possible to generalise in order to relate this variable to a special probability distribution: the Dirac distribution (see Fig. 1.4 for a continuous example of this distribution). In this case,

$$P(X) = \begin{cases} 1, & X = x \\ 0, & X \neq x \end{cases} \quad \Rightarrow H(X) = 0.$$

On the other hand, in the opposite case for which variable X is maximally random, any outcome is likely to occur, and therefore all outcomes are equally probable – we are clearly facing a uniform distribution. Since any value of X is equally probable, the largest possible entropy is obtained,

$$P(X) = 1/M \Rightarrow H(X) = \log M,$$

where M represents the size of the measurable set of X.

Therefore, entropy is a *measure of uncertainty* of the underlying distribution $P(X)$: the more uncertain the distribution, the more information is gathered by observing its value.

Finally, *joint entropy* is defined as the entropy of a joint distribution, given generically by

$$H(X_1, \ldots, X_N) = \sum_{X_1} \cdots \sum_{X_N} \Big\{ -P(X_1 \cap \ldots \cap X_N) \log P(X_1 \cap \ldots \cap X_N) \Big\}. \tag{1.18}$$

Note that, as mentioned above, this particular formulation simplifies significantly if the N random variables $X_1 \ldots X_N$ are independent, as it becomes equal to the sum of the individual entropies.

1.4.2 Mutual Information and Perception

We started this chapter by discussing the importance of dealing with uncertainty in perception; therefore, it seems only natural that we now define how to quantify how much uncertainty about the world state X is decreased by sensing Y. Using the notion of entropy defined previously, this is achieved by taking the difference between the entropy of $P(X)$ and the entropy of $P(X \mid Y)$,

$$I(X;Y) = H(X) - H(X \mid Y), \tag{1.19}$$

where $H(X \mid Y)$ is the *conditional entropy*, yielded by the entropy of the conditional distribution on the world's state $P(X \mid Y = y)$ averaged over the probability of observation $P(Y = y)$,

$$\begin{aligned} H(X \mid Y) &= E_{P(Y)}\left[E_{P(X|Y)}[-\log P(X \mid Y)]\right] \\ &= \sum_Y P(Y) \sum_X -P(X \mid Y) \log P(X \mid Y). \end{aligned} \tag{1.20}$$

This difference, $I(X;Y)$, is called the *mutual information* of X and Y, and it is symmetric with respect to both variables. This can be attested by noting that the entropy of the joint probability, $P(X,Y) = P(Y \mid X)P(X) = P(X \mid Y)P(Y)$, is given by

$$H(X,Y) = H(X) + H(Y \mid X) = H(Y) + H(X \mid Y), \tag{1.21}$$

which means that mutual information can be computed in three different ways:

$$\begin{aligned} I(X,Y) &= H(X) - H(X \mid Y) && (1.22a) \\ &= H(Y) - H(Y \mid X) && (1.22b) \\ &= H(X) + H(Y) - H(X,Y). && (1.22c) \end{aligned}$$

1.4.3 Information Gain – The Kullback-Leibler Divergence

Imagine that we now would wish to compute the *information gain* about a random variable X obtained from an observation that a random variable A on which X is known to depend takes the value $[A = a]$; in other words, the information gain by going from a prior distribution $P(X)$ for X to an updated posterior distribution $P(X \mid a)$ for X given a.

The expected value of the information gain is, in fact, the reduction in the entropy of X achieved by learning the state of the random variable A. Simplifying notation by denoting the posterior distribution as $Q(X) = P(X \mid A)$, we have that the expected value of the information gain is given by the *Kullback-Leibler (KL) divergence*,

$$D_{KL}(Q||P) = E_{Q(X)}\left[\log_2\left(\frac{Q(x)}{P(x)}\right)\right] = \sum_X Q(x)\log_2\left(\frac{Q(x)}{P(x)}\right). \tag{1.23}$$

This expression is commonly used as a measure of the difference between two related distributions: in general $D_{KL} \geq 0$, and it becomes zero if the distributions match exactly. However, it cannot be called a "distance", since it usually does not satisfy the symmetry condition, i.e., most of the times, $D_{KL}(Q||P) \neq D_{KL}(P||Q)$.

In summary, the Kullback-Leibler divergence provides a statistical measure that quantifies in bits how close a probability distribution $P(X)$ is to the model distribution (i.e. the updated plausibility of the hypothesis) $Q(X)$.

1.5 Graphical Models – Bayesian Networks

As can be understood from the previous section, Bayesian inference may involve rather complicated formalisations and algebraic manipulations, rendering the Bayesian approach to modelling a very hard reading exercise. For this reason, in many applications, graphical models have been devised to improve readability and for clarity of exposition.

The most popular graphical representations used for probabilistic modelling (although there are others – see Pearl [5] for more examples) would be Bayesian networks (BNs), which are directed acyclic graphs whose nodes represent variables and the arcs express the dependences between the linked variables.

The Bayesian network constitutes a complete probabilistic model of the variables in a domain, containing sufficient information to answer all probabilistic queries about these variables needed to perform inference [5]. A BN for a set of variables $\mathbf{X} = \{X_1, \ldots X_N\}$ consists of

1. A network structure, the directed acyclic graph S, that encodes the set of conditional independence assumptions about the variables in $\mathbf{X}$.
2. A set of local probability distributions associated with each variable.

Together, these components define the joint probability distribution for $\mathbf{X}$, $P(X_1 \wedge \ldots \wedge X_N)$, and thereby allow the establishment of the desired generative model. The second component means this joint distribution is *decomposable* using the Bayesian inference rules, resulting in local probability distributions with a single variable to the left of the conditioning bar[9].

In particular, in a BN with nodes $X_1 \ldots X_N$, it is possible to calculate the probability of the proposition $[X_i = x_i]$ given the values of all other variables in the domain $[X_k = x_k], k = 1 \ldots N, k \neq i$, i.e. the posterior $P([X_i = x_i] \mid x_1 \wedge \ldots \wedge x_{i-1} \wedge x_{i+1} \wedge \ldots \wedge x_N)$.

An example of a Bayesian network using a common (but not unique) graphical notation is shown in Fig. 1.6, demonstrating how to build from it the model's joint distribution and its corresponding decomposition. Bayesian networks and their usage will be further detailed in Chapter 3.

[9] This is, in fact, the main drawback of traditional Bayesian networks, which do not allow inference with conjunctions on the left of the conditioning bar without the help of an intermediate variable – see Chapter 3 to better understand the implications.

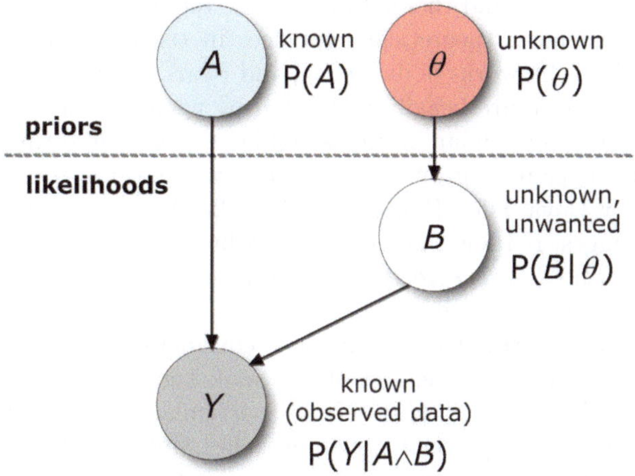

Fig. 1.6. Example of a simple Bayesian network and respective notation. Arrows converging to a node denote a conditional pdf for its variable. *Usually* arrows direct causality, and inversely direct dependency, so, in this case, the conditional pdfs are $P(Y \mid A \wedge B)$ and $P(B \mid \theta)$. Terminal nodes (i.e. nodes with no converging arrows) correspond to prior pdfs, in this case $P(A)$ and $P(\theta)$. The model's joint distribution is constructed by multiplying the product of all priors with the product of all conditional pdfs, a process called *decomposition*, in this case resulting in $P(A \wedge B \wedge \theta \wedge Y) = P(A)P(\theta)P(B \mid \theta)P(Y \mid A \wedge B)$. This representation formalises the step of forward modelling, and consequently allows the application of the subsequent step of inference to determine the *searched* unknown variables (in this case, θ).

1.6 Final Remarks and Further Reading

This chapter intends to be a primer on probabilistic inference, in particular inference in the context of the Bayesian approach; it is *not* a historical account, an epistemological or philosophical essay, nor a complete manual on probabilistic methodologies and notation. Its objective is, in fact, to introduce the essential theoretical foundations which will allow the reader to apply probabilistic approaches while feeling a minimum sense of grasping the "whys, whats and hows" that support these techniques.

The main reference on Bayesian approaches to Robotics and Artificial Intelligence used in this text is the excellent textbook by Bessière, Laugier, and Siegwart [1]. The reader will find a lot of affinities between both texts, although our book contributes differently by concentrating in providing a more thorough and systematic analysis of the application of Bayesian approaches to the specific context of robotic perception. Bessière et al. present a thorough historical overview of seminal references in all fields that one might think of that use probabilistic solutions, which we strongly recommend reading by all who wish to dive deeper into the fascinating world of probabilistic approaches and the philosophy behind them. Two of these references, which we will risk claiming are the main theoretical references in terms of the Bayesian approach, would be the works of Jaynes [4] and Pearl [5]. Bayesian approaches to robotic perception are heavily motivated by what several researchers believe to happen in the human brain, as will be demonstrated throughout this book – for a deeper insight on these matters, we refer the reader to the excellent book by Doya, Ishii, Pouget, and Rao [2]. As for the information and entropy, we suggest reading the seminal work by Shannon [6], freely available online in a reprinted and corrected version. Finally, further reading material on statistical inference can be found in [3].

In the course of writing this chapter, a considerable effort was made to maintain consistency with the rest of the book and also a minimum accordance with cited references. This effort is particularly evident in one of the most varying aspects of probabilistic approaches, *notation*. Many readers are put off when studying probabilistic approaches and solutions by the diverging views on notation used throughout the existing supporting literature, further complicated by the idiosyncrasies introduced by each author when presenting their specific work. With this in mind, the authors ensured that in the first part of this monograph, the notation presented in this chapter would be followed to the letter, and that it would be as faithful as possible to the notation used in the worked out examples of the second part. The notation used in this book closely follows, with very few exceptions in order to keep consistency with other references, what is presented in [1].

References

1. Bessiére, P., Laugier, C., Siegwart, R. (eds.): Probabilistic Reasoning and Decision Making in Sensory-Motor Systems. STAR, vol. 46. Springer, Heidelberg (2008) ISBN 978-3-540-79006-8 22, 23, 35, 36
2. Doya, K., Ishii, S., Pouget, A., Rao, R.P.N. (eds.): Bayesian Brain – Probabilistic Approaches to Neural Coding. MIT Press (2007) ISBN 978-0-262-04238-3 4, 36
3. Berthold, M.R., Hand, D.J. (eds.): Intelligent Data Analysis — An Introduction, 2nd edn. Springer (2003) ISBN 978-3-540-43060-5 36
4. Jaynes, E.T.: Probability Theory: the Logic of Science. Cambridge University Press (2003) 5, 6, 19, 20, 21, 22, 36
5. Pearl, J.: Probabilistic Reasoning in Intelligent Systems: Networks of Plausible Inference, revised second printing edn. Morgan Kaufmann Publishers, Inc., Elsevier (1988) 5, 20, 22, 34, 36
6. Shannon, C.E.: A Mathematical Theory of Communication. The Bell System Technical Journal 27, 379–423, 623–656 (1948), `http://infocom.uniroma1.it/~robby/tic1/shannon1948.pdf` 36

2

Representation of 3D Space and Sensor Modelling within a Probabilistic Framework

Between stimulus and response there is a space. In that space is our power to choose our response. In our response lies our growth and our freedom.
Man's Search for Meaning, *Viktor Frankl (1946)*

For the wise man looks into space and he knows there are no limited dimensions.
Lao Tzu (500-600 BC)

2.1 Introduction

For living organisms, perception can be defined as a set of *cognitive processes*, in the sense that it consists in the processing of sensorial data in order to generate essential information with the purpose of building a coherent and useful representation of the surrounding world. Perception has been paramount for living beings, its importance having propagated from supporting the original primal objective of survival up to the more recent evolutionary purpose of promoting social interaction.

Now, *cognition* (from the latin *cognoscere*, which means "to know", "to conceptualise" or to "recognise"), although always implying a relatively complex process, can be unconscious. This is, in fact, the case in most of the cognitive processes underlying perception in living beings. However, as can be easily understood from this line of thought, being unconscious does not rule out that these processes involve "reasoning" (even if done automatically).

Therefore, when modelling perception, we are in fact establishing how a cognitive system reasons when relating abstract concepts – or *symbols* – with raw sensor data. This problem has been studied for decades now in Artificial Intelligence and Robotics, under many different names, one of the most well known being the "symbol grounding problem" [35; 16]. Thankfully, as was shown in Chapter 1, plausible reasoning and Bayesian inference address incompleteness and uncertainty, that so often make perception as a symbol grounding problem difficult, inherently.

Unfortunately, the world that is to be analysed by a perceptual system is dynamic: not only the intrinsic properties of objects in the environment are susceptible to changes through time, but also their spatial disposition. This means that the world cannot be perceived with a single glance. Consequently, for the raw sensory data arising from observing objects to be robustly

J.F. Ferreira and J. Dias: *Probabilistic Approaches for Robotic Perception*, Springer Tracts in Advanced Robotics 91,
DOI: 10.1007/978-3-319-02006-8_2,

related to symbols, the spatial properties of a perceptual scene must, at least implicitly, also be assigned a representation.

With this in mind, we have structured this chapter as follows: firstly, we will discuss how space might be referenced and represented; next, we will show how to build probabilistic sensor models taking spatial representation explicitly into account; finally, we will briefly address the difference between detecting and recognising objects, and give an inkling of how to address this problem within a probabilistic framework.

2.2 The Reference of Representation – Egocentric vs. Allocentric

Sensor data will most certainly relate to physical entities, each of which placed in specific locations in the observer's surroundings. The most important and immediate associations that humans and other animals make when trying to make sense of the incoming sensory data are precisely spatial associations, since these generally have imminent significance. Firstly, in simple, primal tasks such as navigation, it is not always as important to know *what* is exact nature of the objects one is observing (i.e. if the road one is following is made of tarmacadam or cobblestones, or if an obstacle one has to avoid is a rock or a chair) as it is to know *where* these objects are. On the other hand, categorising the nature an object is more than often related to where the object is located (for example, when we are interested in getting "the third cup from the left").

A *reference frame* allows the representation of locations of entities in space. It is defined by a geometrical *origin* and a set of (ordinarily, but not necessarily, orthogonal) directions, called *axes*, one per each spatial dimension to be represented.

The "reference of spatial representation", as a point of view from which the observer perceives the world, has been extensively studied in several contexts [25; 8; 21; 34]. There are two different broad ways of qualifying reference frames depending on this point of view:

- **Egocentric:** This is related to point-of-view *centred on the observer* (*ego* being the latin word meaning "self"; i.e. dependent of observer pose). Usually directly or indirectly sensor-referred.
- **Allocentric:** This refers to "other than self" (from the greek word *allos*, meaning "other"). In other words, it refers to any abstract or concrete entity other than observer.

The broad classification of allocentric reference frames is then used with different, more specific meanings depending on the author. For example, it is sometimes used to signify "exocentric" or "geocentric", i.e. the so-called "world reference frame", often abbreviated as *wrf*, which is centred in an abstract origin and usually assumed static. In other cases, it is used to designate

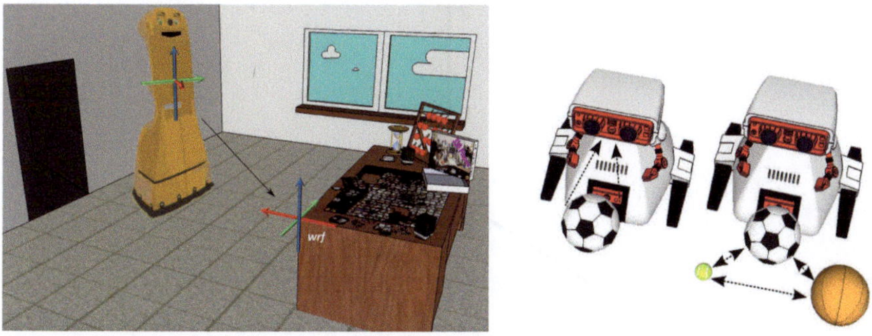

Fig. 2.1. Reference frames. On the left, the relation between the egocentric and geocentric referencing is shown; on the right, the difference between egocentric and allocentric/object-centred referencing is demonstrated.

an "object-centred" reference, i.e. centred on any concrete object [25] – see Fig. 2.1. Allocentric reference frames can also be orientation independent.

How and where the human brain codes information in each of these reference frames [8; 24], how it performs transformations from one frame to another [13; 11; 9], and if there is a common reference frame for central processing or to link perception to action [20] – all these issues have been important subjects of study in cognitive sciences for a long time now. In Chapter 8, an inkling of how the findings of these studies can be used to inspire work in artificial perception will be presented. For now, however, we will only introduce these notions in terms of their mathematical and geometrical relevance for modelling spatial representations.

2.3 Coordinate Systems – Cartesian *vs.* All Others

Reference frames directly associate to another concept: the *coordinate system*, which relates to sets of numbers (the so-called *spatial coordinates*, the order of which inside this set is significant) that uniquely determine positions in space within the reference frame. For simplicity, in many instances the two designations are used interchangeably; however, note that, formally, the coordinate system extends the notion of the reference frame by adding the property of *metric proportion*.

The Cartesian coordinate system, which owes its name to the French mathematician and philosopher René Descartes, is the undisputedly the most commonly used system, relating closely to Euclidean geometry. In this system, coordinate surfaces are orthogonal, and the axes of its reference frame either follow left- or right-handed orientation (the latter being the most frequent) – see Fig. 2.2. Cartesian coordinate systems have widespread use for allocentric mapping, mostly for cultural reasons, and have thus become a *de facto* standard.

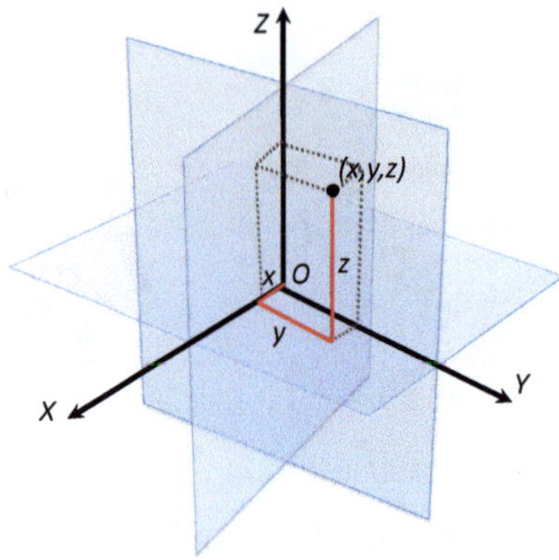

Fig. 2.2. Example of a 3D Cartesian coordinate system, following right-handed orientation

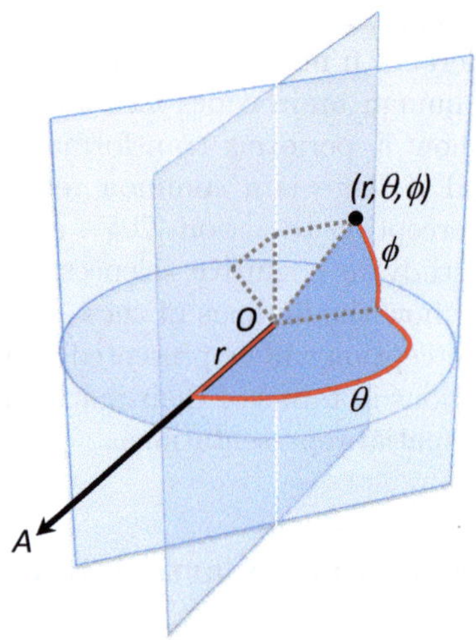

Fig. 2.3. Example of a 3D spherical coordinate system, using distance from origin r, azimuthal angle θ, and elevation angle ϕ as coordinates

Alternative coordinate systems would be polar (2D) and spherical (3D; see Fig. 2.3), which follow many properties of egocentric spatial coding in the brain (more on this in Example 2.1 and later on in Chapter 8), and also cylindrical, parabolic, ellipsoidal, etc.

Coordinate systems generally associate linear scales to each coordinate – this means that distances used in the spatial *representation* of entities are linearly proportional to the *actual* distances in space (i.e. representation is said to be done *to scale*). However, in some cases logarithmic scaling is used

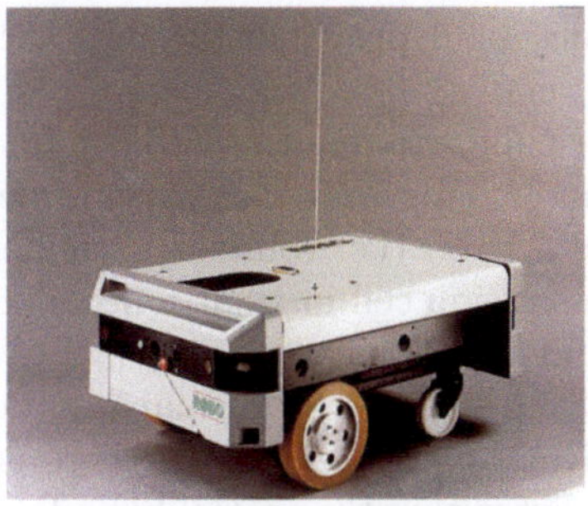

Fig. 2.4. A robot that notably used metric mapping – the robuTER (model shown from circa 1988), by ROBOSOFT, originally designed in INRIA. Used by Leonard and Durrant-Whyte [32] to perform metric mapping using landmarks. (© Dupourque / Wikimedia Commons / CC-BY-SA-3.0 / GFDL.)

– in this case, constant steps along the representation's coordinate axis scale correspond to an exponential step in the real-world environment.

2.4 Mapping to Represent Space

2.4.1 Metric Mapping and Tessellations

We are now in the position to formally introduce the notion of *spatial mapping* as the act of building a representation of one's surroundings in which entities detected by sensors are associated to spatial properties, such as position or velocity.

Arguably the most intuitive and broadly used approach in spatial representation is *metric mapping*. Metric mapping attempts to describe the world in geometric terms, using absolute distances.

Metric mapping is in turn subdivided into two subcategories:

- **Grid-based mapping:** Space is subdivided into a grid of indexable cells, producing a *tesselation.* One or more specific properties are associated to each cell which are then related to raw sensorial data in order to attempt to solve the symbol grounding problem.
- **Feature mapping:** The environment is mapped in terms of geometric primitives, such as lines or corners, or features that can be easily detected and precisely located, the so-called *landmarks.* These primitives are then related to raw sensorial data in order to attempt to solve the symbol grounding problem.

In both cases, coordinate systems and reference frames are obviously crucial in the process of binding symbols to sensorial data. An example of an historical robot that used these approaches is presented in Fig. 2.4.

Perhaps the most successful probabilistic approach for metric mapping has been the *occupancy map or grid*, first introduced by Moravec and Elfes [40; 38; 36]. Consider a cubic volume tesselated into a grid, denoted as $\mathcal{Y}$, of cubic, regular cells, interchangeably indexed and denoted by c. This type of cell, commonly called a *voxel* (i.e. volumetric element), could be replaced by any other type of tesselation cell in what follows (1D, 2D, 3D, nonlinear, nonregular, etc.), with no loss of generality.

The state of each cell in the grid maps as originally defined by Moravec and Elfes is given by its *occupancy* – there is either an object partially or completely present within the spatial boundaries of the cell, in which case the cell is said to be occupied, or the cell is empty. If, instead of just occupancy, more properties are added to the state thus forming a random vector, the occupancy grid generalises to the notion of *inference grid* [36].

In mathematical terms, the occupancy state is represented by the binary random variable O_c, which is 0 when the cell is empty, and 1 otherwise. Since these cell states are mutually exclusive and exhaustive, Axiom 1.5 of Chapter 1 applies, yielding $P([O_c = 0]) + P([O_c = 1]) = 1$. The full state of the occupancy map is given by the conjugation of the states of all the cells composing the grid that represents the environment which is being mapped – formally, $O = \bigwedge_{\mathcal{Y}} O_c$. This defines an *occupancy field*, which is formally a *discrete-state probabilistic process*, while the occupancy grid itself is a *lattice process*, defined over a discrete spatial tesselation.

Temporally, the occupancy grid is traditionally used as a first-order Markov process (see section 1.3.3). Spatially, a similar assumption is commonly made by considering it a *Markov random field* – which is formally defined as a set of random variables having a Markov property described by an undirected graph – of order 0. Put in simple terms, this means that the states of all cells in the grid for each time instant are assumed independent, and thus Bayesian inference can be performed independently for each cell. In any case, computationally more expensive inference can always be implemented in order to model the occupancy grid as a higher-order Markov random field.

In perceptual terms, the occupancy primitive represents a very basic symbol to which raw sensor data can be related to. A deterministic representation of the world can be obtained solely by deciding over the posterior inferred for each cell, for example by assuming the most probable occupancy state. Such a representation for solving the symbol grounding problem makes computations for each cell very inexpensive; on the other hand, it might oversimplify the semantics of the perceived world.

Tesselations used for occupancy grids are usually regular and assume a exocentric Cartesian coordinate system; however, the occupancy map principle holds for any grid configuration, such as tesselations resulting from non-linear division of space, octrees, etc.

Metric mapping has mainly been used in robotics to deal with navigation-based applications, where immediate action is required that doesn't depend on a complex classification of the entities present in the environment.

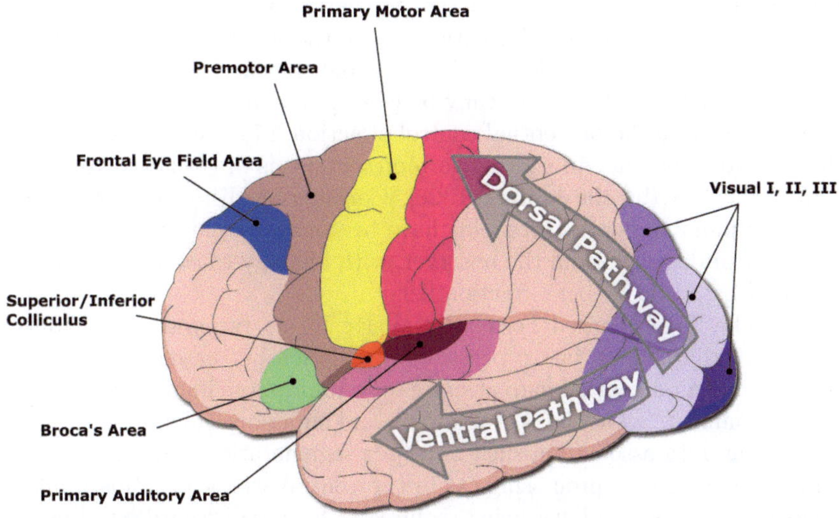

Fig. 2.5. Dorsal and ventral pathways in the human brain. Important sensory sites and saccade and motor areas are also shown, namely the phylogenetically older superior colliculus, which mediates involuntary eye-head movements.

Occupancy grids are very powerful representations, in the sense that an object occupying a given cell can be easily modelled in terms of its effect on sensor data, and that it makes registering measurements from different sensors effortless, in contrast with feature mapping, which of course needs to rely on feature detection and simple data association. Moreover, grid maps are intuitive and formally rigorous.

Conversely, metric maps in general, and occupancy grids in particular, scale ungracefully with size – a great deal of memory and computations are needed to update the representation while using a satisfactory resolution. Occupancy grids have the additional disadvantage of suffering from discretisation issues, such as aliasing and Moiré effects.

In the following text, an example of a specific spatial configuration for an inference grid is presented, demonstrating how important the right choice of an approach to spatial mapping can be.

Example 2.1. **Bioinspired log-spherical inference grid**

Within the human brain, mainly two pathways or streams, anatomically separate albeit interconnected in a complex fashion, have been found to be involved in sensory processing: the *dorsal pathway* and the *ventral pathway*. Both dorsal and ventral systems process information about spatial location, but in very different ways: allocentric spatial information about how objects are laid out in the scene is computed by ventral stream mechanisms, while precise egocentric spatial information about the location of each object in a body-centred frame of reference is computed by the dorsal stream mechanisms, and also the phylogenetically preceding *superior colliculus* (SC), both of which mediate the perceptual control of action [27] – see Fig. 2.5. Finally, direction and distance in egocentric representations are believed to be separately specified by the brain [30; 26]. Considering distance in particular, just-discriminable depth thresholds have been usually plotted as a function of the log of distance from the observer, with analogy to contrast sensitivity functions based on Weber's fraction [29].

These findings inspired Ferreira et al. [1; 10] to propose a probabilistic framework that allows fast processing of multisensory-based perceptual inputs to build a perceptual map of space so as to promote immediate action on the environment (as in the dorsal stream and superior colliculus), effectively postponing data association such as object segmentation and recognition to higher-level stages of processing (as in the ventral stream) — this would be analogous to a tennis player being required to hit a ball regardless of perception of its texture properties. This framework bears an egocentric spherical (i.e. coding 3D distance and direction) spatial configuration, as in Fig. 2.3, and constitutes a short-term perceptual memory performing efficient, lossless compression through log-partitioning of depth.

The proposed framework is thus an inference grid, referred to by the authors as the Bayesian Volumetric Map (BVM). The tesselation of the BVM is primarily defined by its range of azimuth and elevation angles, and by its maximum reach in distance ρ_{Max}, which in turn determines its log-distance base through $b = a^{\frac{\log_a(\rho_{\text{Max}} - \rho_{\text{Min}})}{N}}$, $\forall a \in \mathbb{R}$, where ρ_{Min} defines the *egocentric gap*, for a given number of partitions N, chosen according to application requirements. The BVM space is therefore effectively defined by

$$\mathcal{Y} \equiv \left]\log_b \rho_{\text{Min}}; \log_b \rho_{\text{Max}}\right] \times \left]\theta_{\text{Min}}; \theta_{\text{Max}}\right] \times \left]\phi_{\text{Min}}; \phi_{\text{Max}}\right] \tag{2.1}$$

In practice, the BVM is parametrised so as to cover the full angular range for azimuth and elevation. This configuration virtually delimits a *horopter* for sensor fusion around the egocentric origin $\{\mathcal{E}\}$.

Each BVM cell is defined by two limiting log-distances, $\log_b \rho_{\min}$ and $\log_b \rho_{\max}$, two limiting azimuth angles, $\theta_{\min}$ and $\theta_{\max}$, and two limiting elevation angles, $\phi_{\min}$ and $\phi_{\max}$, through:

$$\mathcal{Y} \supset \mathcal{C} \equiv \left]\log_b \rho_{\min}; \log_b \rho_{\max}\right] \times \left]\theta_{\min}; \theta_{\max}\right] \times \left]\phi_{\min}; \phi_{\max}\right] \tag{2.2}$$

where constant values for log-distance base b, and angular ranges $\Delta\theta = \theta_{\max} - \theta_{\min}$ and $\Delta\phi = \phi_{\max} - \phi_{\min}$, chosen according to application resolution requirements, ensure BVM grid regularity. Finally, each BVM cell is formally *indexed* by the coordinates of its *far corner*, defined as $C = (\log_b \rho_{\max}, \theta_{\max}, \phi_{\max})$.

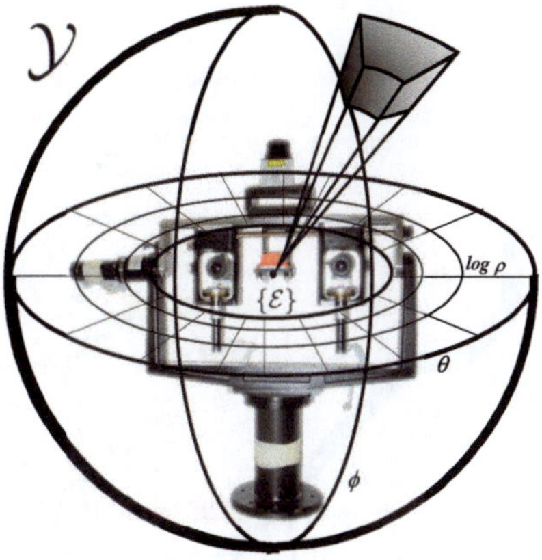

Fig. 2.6. The Bayesian Volumetric Map (BVM) referred to the egocentric coordinate frame of a robotic active perception system

The BVM spatial representation model is shown within its egocentric context in Fig. 2.6. We will build upon this framework and return to it in several examples throughout this manuscript. It presents two advantages when comparing to other representations: (a) an *efficiency advantage*: regular partitioning in Euclidean space, while still manageable in 2D, renders temporal performances impractical in 3D when fully updating a panoramic grid (i.e. performing both prediction/estimation for all cells on the grid) with satisfactory size and resolution (typically grids with much more than a million cells) – additionally, it does so while still accounting for just-discriminable depth thresholds such as those found in visual perception; (b) a *robustness advantage*: the fact that sensor readings are directly referred to in spherical coordinates and consequently no ray-tracing is needed leads to inherent antialiasing, therefore avoiding the Moiré effects which are present in other grid-based solutions in the literature. It can be used for a wide span of robotic applications, especially involving actuation in personal space, where time-to-impact (and therefore distance) is crucial, ranging from active perception and exploration to egocentric navigation and obstacle avoidance.

2.4.2 The Topological Approach

A *topological map* (Fig. 2.7) consists of a graph containing nodes, representing landmarks, and edges connecting them, denoting traversibility – paths or classes of paths, or behaviour sequences for travelling between places [31].

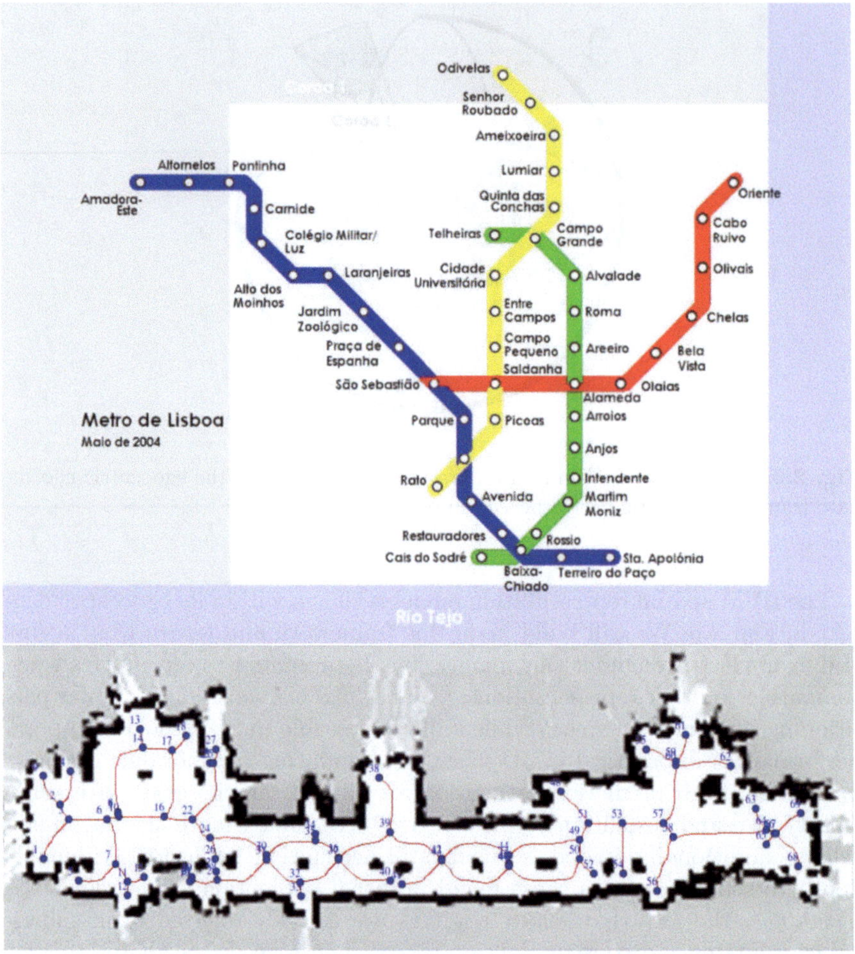

Fig. 2.7. Examples of topological maps. At the top, a 2004 map of the Lisbon Metropolitan. Note that, although relative positioning is roughly maintained, metric scaling is disregarded. At the bottom, an illustrative result of an automatic procedure for extracting topological information from occupancy grid maps for robot navigation, by Portugal and Rocha [2], is shown (reproduced by kind permission).

Landmarks are defined as distinct places in the environment, where its characteristics change significantly. Topological maps are generally used for navigation, and not usually for object representation. Topological maps commonly use an exocentric reference, and inherently dispense a coordinate system.

Topological mapping for navigation includes deciding if a landmark has been previously visited and, if that is the case, when. This is known as the *correspondence problem* in topological mapping [3]. Solving the correspondence problem is non-trivial because many distinct landmarks may appear to be similar to the robot's sensors, a phenomenon designated as *perceptual aliasing*. Consequently, the robot fails to identify the landmark correctly and, as a consequence, to infer the correct topology, resulting in undesirable situations such as not being able to recognise a place it has already passed (i.e. unable to *"close the loop"*).

Many existing techniques approach the mapping problem in a maximum-likelihood framework, with the objective of finding the topology that minimizes some error function [3]. These methods rely on selecting the most likely topology, which can frequently be wrong in the presence of aliasing. In addition, the error function to be optimized may have local minima which also results in an incorrect map.

A common way of overcoming perceptual aliasing involves exploration by the robot until a distinct landmark is observed that allows robot localisation. Also common are approaches that involve behaviour-based control for topological mapping based on exploration. Other works maintain a multiple-hypothesis space over correspondences, which directly relate to posteriors over the set of all possible topologies.

Topological maps offer the advantages of being abstract and therefore compact, and of scaling well with size. However, the difficulty in matching topologies to sensor readings, landmarks or actuator commands, makes the correspondence problem and the consequent loop closing deficiencies a very difficult problem, indeed.

2.4.3 Hybrid and Hierarchical Approaches

Hybrid/hierarchical mapping attempts to capitalise on the complementary advantages of both metric and topological approaches. An example of a robot that notably used this type of method is presented in Fig. 2.8.

One of the trends in hybrid mapping was embodied by the work of Thrun [28]. The approach consisted of extracting a consistent global topological map from a global grid map, which would allow for closing the loop. To construct a 2D grid-based model of the environment, sensor values are interpreted and mapped over time into occupancy states, as described earlier. On top of the grid representation, more compact topological maps are generated by dividing the grid-based map into coherent regions, separated through *critical lines*, which represent narrow passages such as doorways. By partitioning the occupancy map into a small number of areas, the number of topological entities will be several orders of magnitude smaller than the number of cells in the grid representation – the application of this type of approach leads to results similar to the representation at the bottom of Fig. 2.7. However, this approach also inherits two of the disadvantages of grid mapping, namely the considerably larger memory requirements and the necessity for accurate localisation.

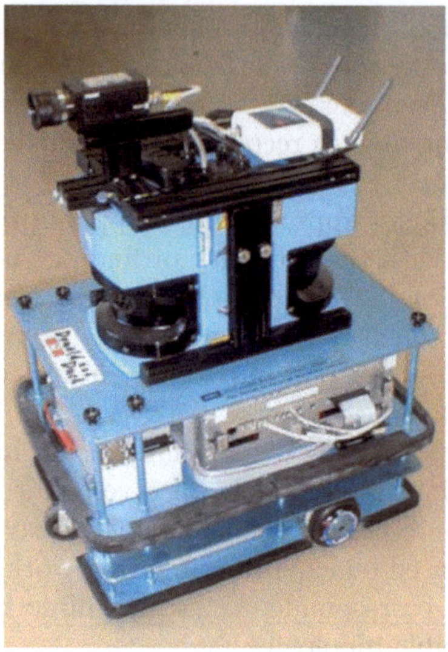

Fig. 2.8. The Donald Duck mobile platform (model shown from circa 2003), one of the notable users of hybrid/hierarchical mapping. This platform is equipped with wheel encoders, a 360° laser range finder and a grey-level CCD camera, used in the work by Tomatis et al. [19]. Reproduced by kind permission.

Conversely, in most other hybrid mapping methods, generally:

- the topological levels are specialised in solving global alignment and navigation problems;
- the metrical levels solve local alignment and object/obstacle problems and represent detailed descriptions of topological map nodes.

One pioneering attempt in this direction was the work of Kuipers and Byun [31]. Their approach would recognise and exploit the qualitative properties of large-scale space before dealing with relatively error-prone geometrical properties. In this approach, there is a control level at the top of the hierarchy, where distinctive places and distinctive travel edges are identified based on the interaction between the robot's control strategies, its sensorimotor system, and the world. The authors defined a *distinctive place* as the local maximum of a distinctiveness measure adequate to its immediate neighbourhood, determined using a hill-climbing control strategy. A distinctive travel edge would analogously be defined through a suitable measure and corresponding path-following control strategy. The following tier, the topological network description, is created by linking the distinctive places and travel edges. Metrical information is then incrementally assimilated into local geometric descriptions of places and edges. Finally, it is merged into a global geometric map. Topological ambiguity arising from indistinguishable places would be resolved at the topological level by the exploration strategy. In a nutshell, the proposed framework is therefore a hierarchical description of the spatial environment, in which a topological network description mediates between a control and a metrical level. With this framework, successful navigation is not critically dependent on the accuracy, or even the existence, of a geometrical description, although a global representation is still maintained in this approach.

A complete departure from global metric representations was promoted by research such as the work of Tomatis, Nourbakhsh, and Siegwart [19]. In this case, metric and topological mapping are completely separated into two levels of abstraction. Metric maps are used only locally for structures (e.g., rooms) that are naturally defined by the environment. For such small spaces, where the drift in the odometry remains uncritical, precise and consistent automatic mapping is perfectly practical. Topological mapping, on the other hand, is used to connect local metric maps. This way, navigation strategies reflects scale-dependent behaviours, namely precise and concrete metric decisions at smaller scales (e.g. obstacle avoidance, moving to a specific spot in a room, etc.) and more abstract decisions at larger scales (e.g., moving from this room to the next, and then turning left). This promotes, according to the authors, compactness of the environment representation and low complexity, allowing for an efficient implementation of the method on a fully autonomous system. Using this approach, the loop is closed in the global topological map based on the information from the topological localisation, while the metric information remains local and does therefore not require further processing, as opposed to the solution proposed by Thrun.

Finally, research such as that presented by Tapus, Battaglia, and Siegwart [14] and Vasudevan, Nguyen, and Siegwart [15] brought a bioinspired point-of-view to the table. Both approaches are hierarchical frameworks with a topological top level in which lower levels are feature-based representations.

The method presented by Tapus et al. [14] was based on the role of place cells in the brain's hippocampus – cells whose firing pattern is dependent on the location of the animal in the environment (through "spatial fingerprints"). The characterisation of the environment using these fingerprints of places is then used within a topological framework. A fingerprint of a place is defined by the authors as a circular list of features, where the ordering of the set matches the relative ordering of the features around the robot. The fingerprint sequence is represented using a list of characters, where each character is the instance of a specific feature defining the signature of a place. In practice, the authors used colour patches and vertical edges extracted from visual information and corners (i.e. extremity of line-segments) derived from a laser-scanner for this purpose.

On the other hand, the probabilistic method introduced by Vasudevan et al. [15] uses object-based representations driving a top-level topological map for indoor mapping, in a process not that dissimilar from what is proposed by Tapus et al.. High-level feature extraction is implemented as an object recognition system, while place identification is implemented using door detection. Note that object recognition here is not implemented in the sense of learning any general properties of the objects in order to categorise or classify them. Its purpose is to transform a set of features, obtained from the object-of-interest using a naive technique, into a robust feature set that incorporates invariance to scale and rotation changes; to a considerable extent it deals with illumination changes and changes in viewing direction as well. Together, object recognition and place identification are encoded to form a hierarchical representation comprising of places, connected by doors and themselves represented by local probabilistic object graphs.

2.5 From Sensation to Perception – The Sensor Model

2.5.1 *Perception as an Ill-Posed Problem*

In blunt formal terms, *sensation* (which one might denote generically as random variable S) is the effect of some phenomenon (which one might denote generically as random variable Φ) on the senses – it thus represents a particular case of a *forward or direct model*, described on Chapter 1 within the context of Bayesian inference. Conversely, perception is the process of *recovering information* on the phenomenon (see 1.4.1 on page 31 in the previous chapter), given the sensation – it thus constitutes the *inverse problem* in this context.

One of the many up-sides of Bayesian inference is that one can formally state very generic and abstract versions of a problem, which can afterwards be made more specific in prototypical fashion. Using this fact to our advantage, we can write out the abstract version of the generative model π of the generic sensation problem as the following decomposition equation [4]

$$P(\Phi \wedge S \wedge \pi) = P(\Phi \mid \pi) P(S \mid \Phi \wedge \pi). \tag{2.3}$$

In this generative model, $P(\Phi \wedge S \wedge \pi)$ is the problem's joint distribution, $P(\Phi \mid \pi)$ is the prior knowledge representing the perceptual system's *expectation on the phenomenon*, and $P(S \mid \Phi \wedge \pi)$ is the *likelihood of the phenomenon* or the *probability of a sensation given the observed phenomenon* – the direct sensor model.

Direct functions are not, in most cases, injective; in other words, the inverse relation is usually not unique. More generally, a problem is said to be:

- *well-posed* when it admits a unique solution – in this case, perception is straightforward;
- *ill-posed* when it admits many solutions – this is the mathematical realisation of the notion of *ambiguity*, and represents, as mentioned on Chapter 1, an important source of uncertainty – or none at all.

Uncertainty due to ambiguity in ill-posed problems is, however, dealt with inherently using Bayesian inference; this will be shown in the following text.

2.5.2 A Solution – Inverting the Problem Using Bayesian Inference

The description of the inverse problem presented previously embodied by the perceptual process is mathematically formalised as $P(\Phi \mid S \wedge \pi)$, i.e. the probability of a specific phenomenon given a sensation.

Using Bayes' rule on equation (2.3), one can infer the inverse problem as being given by

$$P(\Phi \mid S \wedge \pi) = \frac{P(\Phi \mid \pi)P(S \mid \Phi \wedge \pi)}{P(S \mid \pi)},$$

which, by applying the marginalisation rule to the evidence, yields

$$P(\Phi \mid S \wedge \pi) = \frac{P(\Phi \mid \pi)P(S \mid \Phi \wedge \pi)}{\sum_{\Phi} P(\Phi \mid \pi)P(S \mid \Phi \wedge \pi)} \propto P(\Phi \mid \pi)P(S \mid \Phi \wedge \pi).$$

Therefore, given that generative models, by definition, incorporate full probabilistic descriptions of each variable as probability distributions, the generic model for perception includes *full information* on perceptual ambiguities, experienced as multimodal distributions on the likelihood $P(S \mid \Phi \wedge \pi)$ (i.e. distributions that allow more than one phenomenon to have the same probability of having resulted in a specific sensation – a non-injective function), which are perfectly acceptable within the probabilistic framework.

Moreover, if the likelihood $P(S \mid \Phi \wedge \pi)$ is multimodal (even if, in an extreme case, it is uniform, representing the case when no meaningful sensation is retrieved from the sensors), the expectation on the phenomenon expressed by a non-uniform prior helps to *disambiguate* perception, making Bayesian inference a particularly powerful tool. As a matter of fact, this is one of the main strengths of the use of Bayesian approaches for artificial perception.

Additionally, it offers an elegant explanation for many reactions exhibited by natural perceptual systems in the face of ambiguity and uncertainty, such as *explaining away*, in which an *expected cause* of a phenomenon under observation reduces the need to invoke alternative causes. It provides therefore an inkling to how the human brain is able to so easily and effectively perform perceptual decisions in the face of natural scenes (which to most artificial systems would be terribly ambiguous), and on the other hand so prone to failure in the face of unfamiliar or unnatural scenes, giving rise to the so-called *perceptual illusions* – see Fig. 2.9 for examples. Hence the controversial suggestion by some researchers that plausible reasoning, in its formal sense, might be implemented in the biological brain [12].

2.5.3 Dealing with Sensor Fusion

Consider now the case in which we have M sensors observing the phenomenon, yielding $N > M$ sensations, denoted as $\{S_1, \cdots S_N\}$; Equation (2.3) then becomes

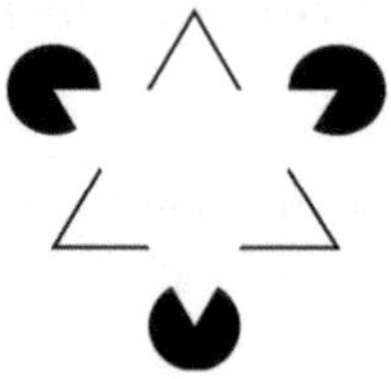

(a) The Kanizsa triangle is an optical illusion first described by the Italian psychologist Gaetano Kanizsa in 1955. In the image above, a white equilateral triangle is perceived, but in fact none is drawn.

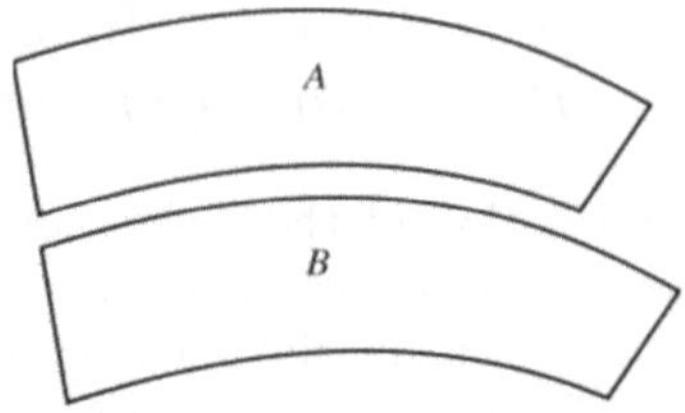

(b) The Jastrow illusion (by American psychologist Joseph Jastrow, 1889). In this illustration, the two figures are identical, although the lower one appears to be larger.

(c) Rubin's vase – a classical example of multistable perception. The brain makes figure-ground distinction between the vase in the centre (usually recognised first) and the contour of the two faces on each side.

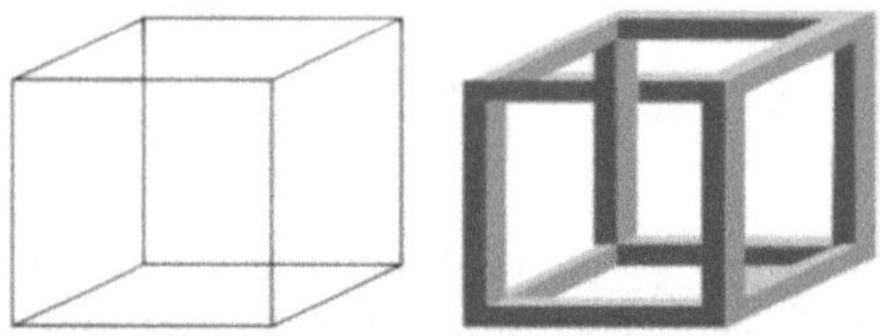

(d) The Necker cube (wire-frame drawing on the left) is an ambiguous drawing of a cube in isometric perspective. When two edges cross, the picture does not show which is in front and which is behind, making the picture ambiguous. When a person stares at the picture, it will often seem to flip back and forth between the two valid interpretations (multistable perception). On the right, inspired on this idea, the "impossible cube".

(e) The image shows what appears to be a black and white checkerboard with a green cylinder resting on it that casts a shadow diagonally across the middle of the board – the squares are actually different shades of grey. The "white" squares in the shadow, one of which is labelled B, are actually the exact same grey value as the "black" squares outside the shadow, one of which is labelled A. The two squares A and B appear very different as a result of the illusion. When interpreted as a 3-dimensional scene, our visual system immediately estimates a lighting vector and uses this to judge the property of the material. (© 1995 Edward H. Adelson, reproducible and distributible freely.)

Fig. 2.9. Several examples of famous visual illusions. Perceptual illusions are common across all senses – the continuity and McGurk auditory illusions, the "cutaneous rabbit" tactile illusion, etc.

$$P(\Phi \wedge S_1 \wedge \cdots \wedge S_N \wedge \pi) = P(\Phi \mid \pi)P(S_1 \wedge \cdots \wedge S_N \mid \Phi \wedge \pi). \qquad (2.4)$$

In general, a robot designer only has the option of modelling each sensor separately, and wants the perceptual system he or she is designing to arrive at a coherent percept from each of the individual sensations. So, how does one go from the individual sensor models/sensations $P(S_i \mid \Phi \wedge \pi)$ to the composite sensor model which constitutes the likelihood of Equation (2.4)?

The solution to this crucial problem in modelling perception comes in the form of *conditional independence* – if it is assumed that two random variables have no immediate or *direct* conditional dependence on each other, they are said to be conditionally independent.

At first glance, most robot designers with practical experience would say that declaring simultaneous readings coming from separate sensors, or coming from an array-type sensor – such as in the case of pixel intensity values from a camera –, or even consecutive readings yielded from the same sensor, as independent is a very coarse approximation. Indeed, sensors affect each other through proximity interference (e.g. heat or electromagnetic effects) through long-distance interference (e.g. in the case of the use of active sensors, such as laser rangefinders or sonars), or both! On the other hand, these sensors might share dependences on a considerable amount of latent variables representing sources of "sensory noise"...

However, as discussed on Chapter 1, these approximations and the uncertainty they introduce are explicitly dealt with in Bayesian modelling, by means of the hidden context variable π, which summarises all of these factors. So, in a sense, the Bayesian approach always attempts to achieve the best possible result from a given model's underlying incompleteness.

Therefore, if the respective (conditional) independence assumptions are **carefully stated**, equation (2.4) can be rewritten as

$$P(\Phi \wedge S_1 \wedge \cdots \wedge S_N \wedge \pi) = P(\Phi \mid \pi) \prod_{i=1}^{N} P(S_i \mid \Phi \wedge \pi), \qquad (2.5)$$

considering all N measurements as independent from one another.

Equation (2.5) represents what is known as *naïve sensor fusion*, due to the approximations introduced by the independence assumptions. Some of the naïveness of sensor fusion using Bayesian inference can be removed by constructing more complicated models with intermediate variables; in any case, conditional independence is an essential tool to make Bayesian models of perception tractable.

Equation (2.5) can be further generalised to include the accumulation of evidence through the collection of sensor readings of a phenomenon taken at consecutive time steps; denoting the conjunction of all sensor measurements as S and each discrete time-step as leading superscripts, and considering that the time-frame under analysis ranges from the initial instant 0 to the present instant T, the expression becomes

$$P(\Phi \wedge S \wedge \pi) = P(\Phi \mid \pi) \prod_{t=0}^{T} \prod_{i=1}^{{}^tN} P({}^tS_i \mid \Phi \wedge \pi). \tag{2.6}$$

Evaluating this expression in every time step is infeasible, as the terms of the products that represent the fusion of likelihoods increase in number exponentially with the passing of time. However, as was discussed in section 1.3.3, if the Markov property is assumed to hold, Bayesian inference can be used recursively, by taking the posterior of the previous time instant as the prior for the present time instant, thus updating the resulting percept with incoming evidence. This is formalised for a specific time instant t as

$$P(\Phi \wedge {}^tS \wedge {}^t\pi) = P(\Phi \mid {}^t\pi) \prod_{i=1}^{{}^tN} P({}^tS_i \mid \Phi \wedge {}^t\pi),$$
$$\text{with} \begin{cases} P(\Phi \mid {}^t\pi) \equiv \text{initial expectation on phenomenon}, & t = 0, \\ P(\Phi \mid {}^t\pi) = P(\Phi \mid {}^{t-1}S \wedge {}^{t-1}\pi), & t > 0. \end{cases} \tag{2.7}$$

Analogously to simultaneous sensor fusion, this process is called *naïve Bayesian update*. It is the simplest of all forms of *dynamic probabilistic loops*.

This kind of update is perfectly acceptable in most applications; however, to model dynamic systems with greater detail (for example, in cases when the phenomenon is itself a dynamic process), it may be insufficient. In Chapter 3, we will present more powerful examples of probabilistic loops to deal with such situations.

2.5.4 Getting in the Right Frame of Mind – Generative vs. Discriminative Models

One of the authors has consistently witnessed fellow researchers, when attempting to use probabilistic approaches to robotic perception for the first time, resisting to use generative models due to a traditional logic (i.e. deterministic) frame of mind. In fact, they would almost invariably strive to directly model the hypothesis conditionally on the data, thereby attempting to find a direct route to obtain the inverse model $P(\text{hypothesis} \mid \text{data})$. In other words, they would try to directly establish conditional models that would obtain the probability of a certain cause from the effect. These are called *discriminative models* and, although they do use probability and probability distributions, they are *not* Bayesian approaches – there is no need to apply Bayes' rule, and no inference is performed.

Fig. 2.10. Generative models viewed as the analogous of a crystal ball for robotic perception. First, one feeds the crystal ball with the robot's knowledge on "This is how my sensor reacts to n well-known stimuli.". From then on, the robot automatically asks the generative model within the crystal ball "What should be my perception, considering this sensation?" by applying Bayes' rule.

So *why* should one use Bayesian modelling to model perceptual processes? This discriminative model approach is not without its merits – for example, no need for computationally expensive inference algorithms means guaranteed real-time performance! However, it defeats the concept of a model "summarising exceptions", the purpose of Fig. 1.1. In fact, the caveat of discrimative models is that they can lead to the undesired bad habit of making too many simplifying assumptions to get to the inverse model directly, generally assuming ill-posed problems as well-posed, therefore risking obtaining little more than a "fancy version of a deterministic model". As a matter of fact, this is a very common criticism to many probabilistic models for robotic perception.

Generative models, on the other hand, *are not stated as conditional models*. Rather, they describe the joint distribution of hypotheses and data in a forward modelling manner, therefore expressing prior assumptions on the degree of plausibility of the generation of those data given each hypothesis (hence the name). However, through the application of Bayes' rule, as was seen in Chapter 1, the solution to the inverse problem (in fact, to any problem stated about any of the involved random variables/propositions) can be inferred from the generative model as a *conditional question* (i.e., "What is $P(\text{hypothesis} \mid \text{data})$?). Nevertheless, the generative model itself is not interchangeable with the questions asked, and is in fact broader in its scope than the latter (and hence its ability to "summarise exceptions").

Given the above, *how* should one think about the perceptual problem to successfully arrive to a generative model?

One can say that the "Bayesian modelling frame of mind", in practice, involves preparing for the implementation of the following steps (Fig. 2.10):

- one first establishes the generative model considering what one knows about the involved (direct) sensation models;
- the robot's perceptual system then becomes enabled to ask the generative model what its estimate is for the inverse model for perception.

The purpose of this book is precisely to provide an introduction and to serve as a reference text for the tools needed to execute these steps. Probabilistic approaches to robotic perception in this sense require a careful assessment of the problem at hand and subsequent statement of the respective generative model, always considering the correct choice of algorithms for efficient, real-time implementation. In particular, forward modelling must be performed either by using preprogramming by the roboticist, or by using suitable learning processes (introduced in Chapter 6), and robotic perception is consequently attained by using the appropriate inference and decision algorithms (introduced in Chapters 3 and 5, respectively).

2.5.5 Examples

In the following text, several published examples of sensor models and their application in robotic perception will be presented and discussed, providing a link between the spatial references and representations described in previous sections with the Bayesian framework for perception defined in this section.

Example 2.2. **Sensor model for in-hand exploration of objects**

Faria, Martins, Lobo, and Dias presented in [5] a representation of 3D object shape using a probabilistic object-centred, volumetric, and Euclidean occupancy grid, the state of which would be estimated in the course of in-hand exploration.

The exploratory procedure was based on contour following through the recording of 3D positions of fingertips while stroking an object's surface, using readings provided by a Polhemus Liberty magnetic tracker sensors attached to the thumb, index, middle fingers. Each sensor returns the 3D coordinates based on sensors' frame of reference. The frame rate for each sensor is 15 Hz, and the mean sensing error is of 2 mm, according to the manufacturer.

Following a long (but not unique) tradition in sensor model notation of representing random variables relating to distance-based sensor readings as

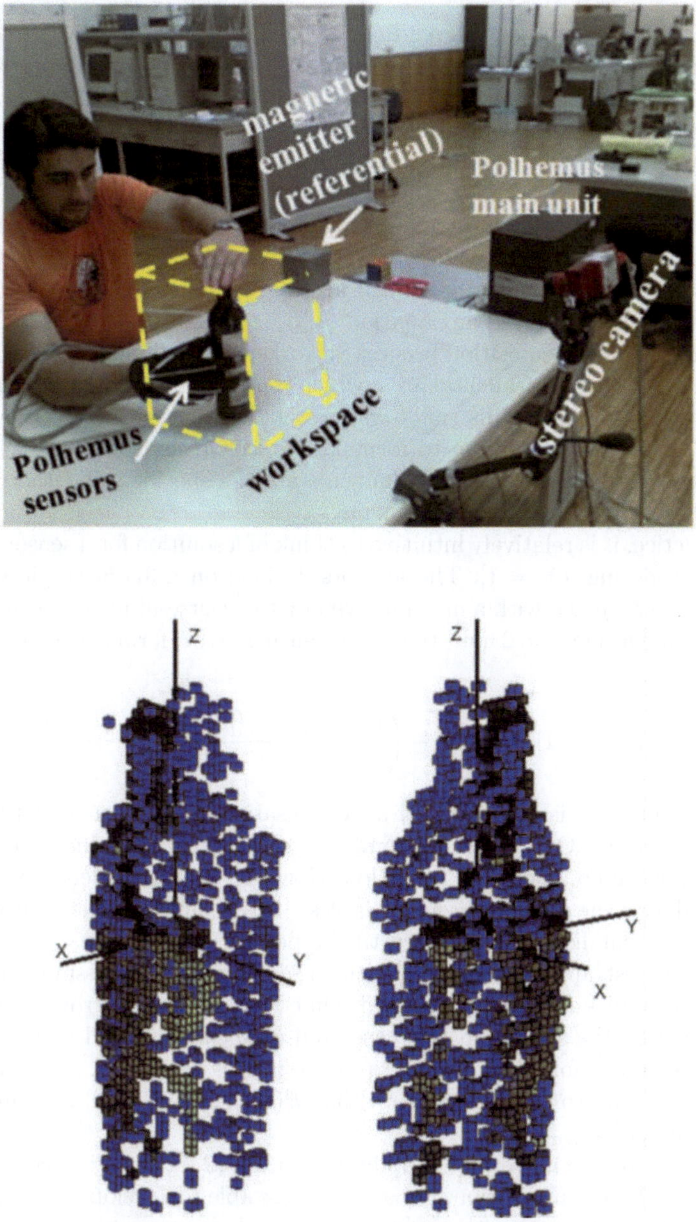

Fig. 2.11. Experimental setup (top) and some results showing voxels with estimated $P([O_C = 1]) > .8$ in an object-centric reference frame (bottom) of the research work of Faria, Martins, Lobo, and Dias [5]. Reproduced by kind permission.

Z, in this particular case the readings yielded by the magnetic tracker will be denoted as $Z_{grasp} = [x, y, z]$ (for simplicity and by abuse of notation, no specific distinction is being made here between readings from different sensors or time-instants), the conjunction of which will replace the generic random variable S in its role of symbolising sensation in the generative model given by Equation (2.3).

Similarly, since a simple occupancy grid as presented in section 2.4.1 is being used, the phenomenon being observed by the magnetic tracker is denoted as O_c, where c represents a given cell in the grid $\mathcal{M}$, taking the role of the generic random variable $\varPhi$ in Equation (2.3).

During the data acquisition process, a workspace of 35 cm × 35 cm × 35 cm was defined in the experimental area, which represented the outer boundaries of the occupancy grid. This space was tesselated into .5 cm × .5 cm × .5 cm voxels. The authors had then to formulate a sensor model $P(Z_{Grasp} \mid O_C)$, that would describe the effect of a known state of a specific cell, either $[O_C = 0]$ or $[O_C = 1]$, on sensor measurements.

In practice, it is relatively intuitive to think of a solution for a sensor model when considering $[O_c = 1]$. The authors decided on a 3D isotropic normal distribution $\mathcal{N}(\mu, \sigma)$, with a mean μ given by the centre-of-mass of cell C and standard deviation $\sigma = 2\,\text{mm}$, the mean sensing error, formalised as

$$P(Z_{Grasp} \mid [O_c = 1]) = \exp\left(-\left(\frac{(x-\mu_x)^2 + (y-\mu_y)^2 + (z-\mu_z)^2}{2\sigma^2}\right)\right).$$

Not as intuitive is the choice of a formalisation of a sensor model for the case of $[O_c = 0]$. One would be tempted to imagine that, since the cell is empty, there is no associated sensation. However, sensor models are subject to error due to the effect of latent variables – this is, after all, the main reason for using probabilistic models for artificial perception.

The simplest approach to deal with misdetections is to assume that we have no way of knowing before hand what would be the (erroneous) reading yielded by the sensors in this case, and as such we should attribute the same probability for *any* reading to occur in this situation. Therefore, we can assume (and so did the authors) that $P(Z_{Grasp} \mid [O_c = 0])$ is a uniform distribution, $\mathcal{U}(Z_{Grasp})$.

Resorting to the naïve sensor update equation (2.7), applying Bayes' rule together with marginalisation, Faria et al. were able to develop an simple, yet powerful model to deal with the uncertainty inherent to object reconstruction via in-hand exploration – the corresponding experimental setup and an example of results are shown on Fig. 2.11.

We will build upon this model and return to it in several examples later on in this manuscript.

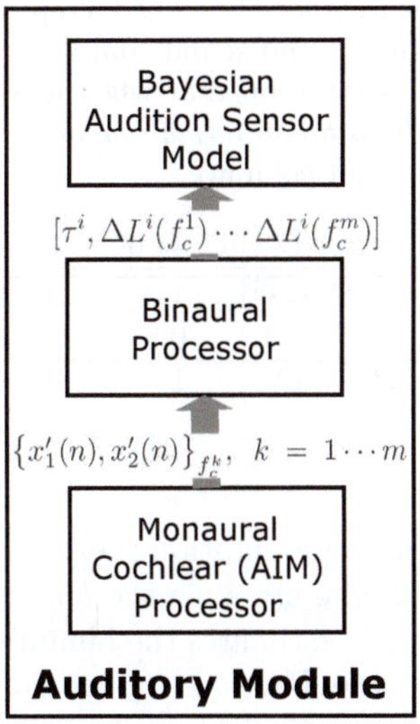

Fig. 2.12. The IMPEP Bayesian binaural system

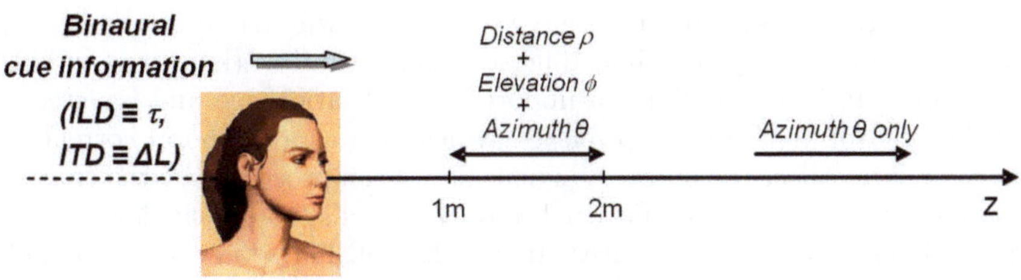

Fig. 2.13. Using binaural cues for 3D localisation of sound-producing objects. Within 2 meters range, the intersection of the interaural level difference (ILD) and interaural time difference (ITD) volumes allows for the full 3D localisation of a sound source. If the source is more than 2 meters away, the change in ILD with source position is too gradual to provide spatial information (at least for an acoustically transparent head), and the source can only be localised in azimuth. In this case, binaural cues in biological binaural-based perception, in humans and other mammals, are complemented by monaural cues resulting from audio filtering by the outer ear (*pinnae*).

Example 2.3. **Sensor model for binaural sensing**

Ferreira, Pinho, and Dias presented in [7] a Bayesian binaural framework composed of three distinct and consecutive processors (Fig. 2.12): the *monaural cochlear unit*, which processes the pair of monaural signals $\{x_1, x_2\}$ coming from the binaural audio transducer system by simulating the human cochlea,

Table 2.1. Probability table used for $P(S_c \mid O_c)$, empirically chosen so as to reflect the indisputable fact that there is no sound source in a cell that is not occupied (left column), and the safe assumption that when a cell is known to be occupied there is little way of telling *from this information alone* if it is in this condition due to a sonorous object or not (right column)

$P(S_c \mid O_c)$	$[O_c = 0]$	$[O_c = 1]$
$[S_c = 0]$	1	.5
$[S_c = 1]$	0	.5
$\sum P(s_c \mid O_c)$	1	1

so as to achieve a *tonotopic* representation (i.e. a frequency band decomposition) of the left and right audio streams; the *binaural unit*, which correlates these signals and consequently estimates the binaural cues and segments each sound source; and, finally, the *Bayesian 3D sound source localisation unit*, which applies a Bayesian sensor model so as to perform localisation of sound-sources in 3D space. This framework was used in a robotic active perception system, with two AKG Acoustics C417 linear microphones and an FA-66 Firewire Audio Capture interface from Edirol.

Sound waves arising from a source on our left will arrive at the left ear first. This small, but perceptible, difference in arrival time (known as an ITD, interaural time difference) is an important localisation cue and is detected by the *inferior colliculus* in primates, which acts as a temporal correlation detector array, after the auditory signals have been processed by the cochlea. Similarly, for intensity, the far ear lies in the head's "sound shadow", giving rise to interaural level differences (ILDs) [22; 18]. ITDs vary systematically with the angle of incidence of the sound wave relative to the interaural axis, and are virtually independent of frequency, representing the most important localisation cue for low frequency signals ($< 1500\,\text{Hz}$ in humans). ILDs are more complex than ITDs in that they vary much more with sound frequency. Low-frequency sounds travel easily around the head, producing negligible ILDs. ILD values produced at higher frequencies are larger, and are increasingly influenced by the filter properties of each external ear, which imposes peaks and notches on the sound spectrum reaching the eardrum.

Moreover, when considering sound sources within $1-2$ meters of the listener, binaural cues alone can even be used to fully localise the source in 3D space (i.e. azimuth, elevation and distance). Iso-ITD surfaces form hollow cones of confusion with a specific thickness extending from each ear in a symmetrical configuration relatively to the medial plane. On the contrary, iso-ILD surfaces, which are spherical surfaces, delimit hollow spherical volumes, symmetrically placed about the medial plane and centred on a point on the interaural axis [23]. Thus, for sources within 2 meters range, the intersection of the ILD and ITD volumes is a torus-shaped volume [23]. If the source is more than 2 meters away, the change in ILD with source position is too gradual to provide spatial information (at least for an acoustically transparent head), and the source can only be localised inside a volume within the cone of confusion delimited by the respective iso-ITD surfaces [23] – see Fig. 2.13.

Given this background, Ferreira et al. decided to adapt the solution by Faller and Merimaa [17] to implement the binaural processor. Using this algorithm, interaural time difference and interaural level difference cues are only considered at time instants when only the direct sound of a specific source has nonnegligible energy in the critical band and, thus, when the evoked ITD and ILD represent the direction of that source (corresponding to the process involving the *superior olivary complex* (SOC) and the *central nucleus of the inferior colliculus* (ICc) in mammals). They show how to identify such time instants as a function of the *interaural coherence* (IC). The source localisation suggested by the selected ITD and ILD cues are shown to imply the results of a number of published psychophysical studies related to source localisation in the presence of distractors, as well as in precedence effect conditions [39]. This algorithm thus amplifies the signal-to-noise ratio and facilitates auditory scene analysis for multiple auditory object tracking.

Faller and Merimaa's cue selection method, as the authors point out, can be seen as a "multiple looks" approach for localisation, which provides the motivation for our implementation. Multiple looks have been previously proposed to explain monaural detection and discrimination performance with increasing signal duration [33]. The idea is that the auditory system has a short-term memory of "looks" at the signal, which can be accessed and processed selectively. In the context of localisation, the looks would consist of momentary ITD, ILD, and IC cues. With an overview of a set of recent cues, ITDs and ILDs corresponding to high IC values are adaptively selected and used to build a histogram that provides a statistical description of gathered cues (see Fig. 2.14).

Finally, the binaural processor capitalises on the multiple looks configuration and implements a simple auditory scene analysis algorithm for detection and extraction of important auditory features to build conspicuity maps and ultimately a saliency map, thus providing a functionality similar to the role of the *external nucleus of the inferior colliculus* (ICx) in the mammalian brain. The first stage of this algorithm deals with figure-ground (i.e. foreground-background) segregation and signal-to-noise ratio. In signal processing, the energy of a discrete-time signal $x(n)$ is given by [37]

$$E = \sum_{-\infty}^{\infty} |x(n)|^2$$

Using this notion, a simple strategy can be followed to selectively apply the multiple looks approach to a binaural audio signal buffer so that only relevant audio snippets are analysed. This strategy goes as follows: given a binaural signal buffer of N samples represented by the tuple $\{x'_1(n), x'_2(n)\}$, the average of the energies of the component signals $x'_1(n)$ and $x'_2(n)$ is

$$E_{avg} = \frac{\sum_1^N |x'_1(n)|^2 + \sum_1^N |x'_2(n)|^2}{2} \tag{2.8}$$

and can be used as a noise gate so that only when $E_{avg} > E_0$ ITDs, ILDs and ICs triplets are collected for the buffer, yielding multiple looks values only for relevant signals (just the ITD-ILD pairs corresponding to high IC values are kept in conspicuity maps per frequency channel), while every other buffer instantiation is labelled as irrelevant noise. E_0 can be fixed to a reasonable empirical value or be adaptive, as seems to happen with human hearing. A set of results exemplifying this algorithm is presented on Fig 2.15.

Once the multiple looks information is gathered, since ITDs are proven to be stable across frequencies for a specific sound source at a given azimuth regardless of range or elevation, the ITD conspicuity maps may be summed over all frequencies, in a process similar to what is believed to occur in the ICx, in computational terms known as a *summary cross-correlogram* (again see Fig. 2.14). From the resulting one-dimensional signal, the highest peaks may be taken as having been effected by the most important sound-sources represented in the auditory image. Then, a search is made across each frequency band to find the closest ITD and its ILD pair, for each reference ITD, thus building n-sized vectors (for $m = n - 1$ frequency channels) for each relevant sound source of the form

$$Z = [\tau, \Delta L(f_c^1) \cdots \Delta L(f_c^m)] \tag{2.9}$$

with τ representing the ITD resulting from the highest peak of the summary cross-correlogram, and $\Delta L(f_c^1) \cdots \Delta L(f_c^m)$ representing the corresponding ILDs for each frequency band.

A sensor model is then used by the Bayesian 3D sound source localisation unit devised by Ferreira et al. to determine the location of objects in space in terms of occupancy on the BVM, the log-spherical inference grid described on Example 2.1. In order to define such a sensor model from which to infer $P(O_c \mid Z)$, using Z as defined on (2.9), a little more work is needed than in the previous example.

First, one needs note the fact that, while all objects occupy space (in particular sound-producing objects), not all objects produce sound at a given instant. Ferreira et al. therefore introduced an intermediate binary random variable, S_c, that only signals occupancy resulting from sound sources. Therefore, a simple probability distribution can be defined relating S_c to O_c, given in Table 2.1. Then, Ferreira et al. acknowledged that the ITD and ILDs can be assumed as conditionally independent from one another – their common relation is reflected by their dependence on spatial location through S_c. This is, of course, an approximation, since the computation of ILDs depends on establishing an ITD first. However, the authors proved through the results presented in [7] that this approximation (which introduces uncertainty) is acceptable and dealt with appropriately by the probabilistic nature of the model.

Consequently, the following joint decomposition equation is formed

$$P(Z \wedge S_c \wedge O_c) = P(O_c) \underbrace{P(S_c \mid O_c) P(\tau \mid S_c) \prod_{k=1}^{m} P(\Delta L(f_c^k) \mid S_c)}_{\text{Gives } P(Z \mid O_c) \text{ through } \sum_{S_c}}, \quad (2.10)$$

which, through the marginalisation of S_c, gives

$$P(Z \wedge O_c) = P(O_c) P(Z \mid O_c),$$

the common formulation for occupancy grids, from which $P(O_c \mid Z)$ can be obtained through Bayesian inference.

Finally, Ferreira et al. defined the types of distribution of the likelihoods for the sensor models, $P(\tau \mid S_c)$ and $P(\Delta L(f_c^k) \mid S_c)$, for both $S_c = 1$ and $S_c = 0$, and how to obtain them (a process commonly called *identification* – see Chapter 3). This process will be described in Chapter 6.

Results of applying these models are presented on Figs. 2.16 and 2.17.

2.6 To Detect or to Recognise? – Wrapping It Up

All sensor models presented as examples until now have been dedicated to *detecting* objects, while simultaneously localising them in space. However, perception as a symbol grounding mechanism has been ubiquitously accepted as being a process leading beyond simple detection towards object *recognition*, therefore more completely describing the semantics and context of the sensed environment.

In fact, the generative equation (2.3) and all its evolution up until (2.7) represent the *probability model* of what is called a *Bayesian classifier*. Bayesian classifiers replace the notion of sensation, using these same equations, by the more generic notion of *features*, which can range from raw sensations to already a result of higher-level perceptual processes; on the other hand, the perceptual variable Φ for a Bayesian classifier is related to a discrete set of *classes*. As a matter of fact, the occupancy grid is a special case of a Bayesian classifier applied to each cell on the tesselation, for which only two classes are considered – either occupied or empty. One can say, consequently, that detection is the simplest case of classification.

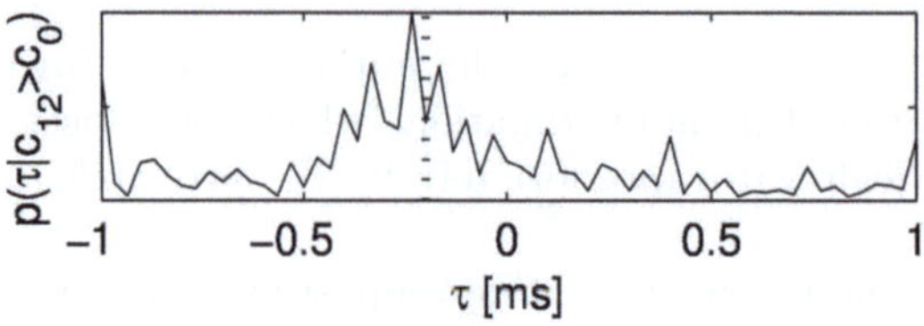

Fig. 2.14. Example of the use of an adaptation of the cue selection method proposed by [17] using a 1 s "multiple looks" buffer. Represented in the figure is a histogram of collected ITD cues (τ) corresponding to high IC levels ($c_{12} > c_0$) for a particular frequency channel of a 1 s audio snippet. This histogram is interpreted as a distribution corresponding to the probability of the occurrence of ITD readings, which is then used as a conspicuity map in order to perform a *summary cross-correlogram* over all frequencies (see main text for more details).

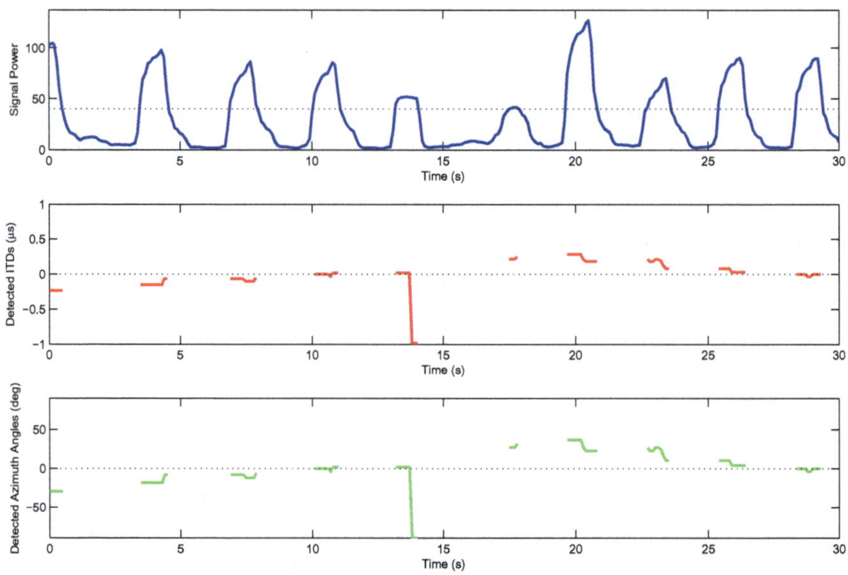

Fig. 2.15. Binaural processing results of an approximately 30 second-long audio snippet of a typical "cocktail party" scenario, with the main voice repeatidely calling out "Nicole, look at me" approximately every 3 s, while other voices can be heard coming from sites close to the robotic head, elsewhere in the lab. The active perception head was moved while the main speaker was kept still, first keeping the speaker to the right and slowly travelling towards the centre, then keeping the speaker to the left and again slowly moving towards the centre. Top — the effect of the signal power-based figure-ground segregation noise gate is shown (dashed line represents gate threshold); Middle — ITD estimates for the most salient sound; Bottom — corresponding azimuth estimates. These results show the performance of the binaural processor under difficult conditions, the only "failure" being the estimates corresponding to the 14 s instant: for a signal power above the interest threshold, the background noise (i.e., some other voice in the lab) was more salient than the main voice.

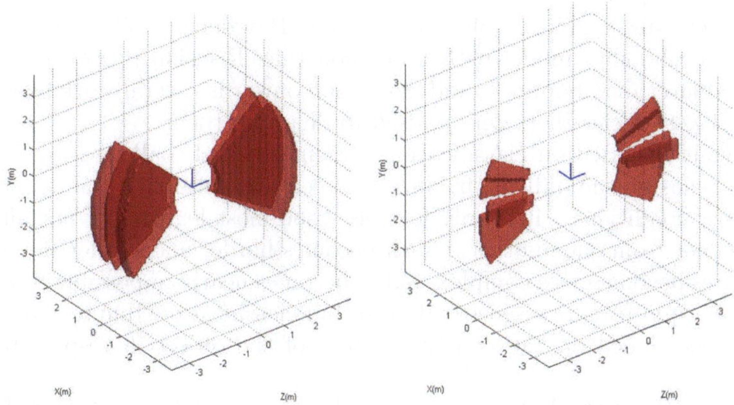

Fig. 2.16. Inference results for the processing of an audio snippet of a human speaker placed in front of the binaural perception system. Cells within the log-spherical sensor-space with probabilities of occupancy greater than .75 are depicted in red, and the egocentric referential in blue (X-axis, Y-axis and Z-axis indicate right-to-left, upward and forward directions, respectively). On the left, result of inference using ITDs only; on the right, result of adding ILDs: note the effects on distance and elevation.

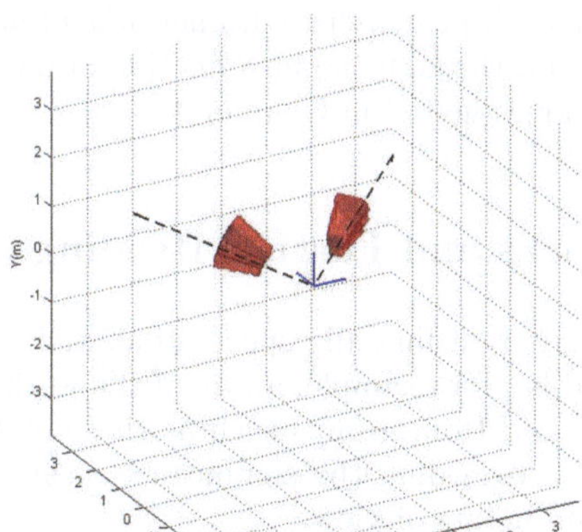

Fig. 2.17. Inference results for the processing of an audio snippet of a sound-source placed at $\rho = 1320\,\text{mm}$, $\theta = 36°$, $\phi = 20°$. All that is depicted has the same meaning as in Fig. 2.16; two dashed directional lines at (θ, ϕ) and $(180° - \theta, \phi)$ have been additionally plotted to demonstrate the effect of front-to-back confusion. This phenomenon can be countered in two different ways: either by rotating the perceptual system, causing the occupancy probabilities of the correct cells to be confirmed and of incorrect cells to be decreased by accumulated evidence with subsequent inference steps, respectively, or by using artificial *pinnae* so as to enforce asymmetry in the HRTF readings and performing calibration using a half-sphere instead of only a quadrant. The fact that $\theta >> 0°$ means that precision in elevation and distance is improved as compared to Fig. 2.16.

However, this is an incomplete definition of Bayesian classifiers: they combine the probability model with a decision rule. Decision processes will be discussed later, in Chapter 5, and from then onwards several working examples of Bayesian classifiers will be used to illustrate them.

Recognition based on classification presents several challenges to a modeller. It relies on the difficult balance between computability and invariance to sensory conditions – it is only feasible when most features (if not all) are independent from one another, which becomes harder and harder to control as features become more complex, but it becomes nearly impossible to model in a single step from raw sensory data to full classification, due to their variability conditioned on the relative context of the object in respect to the sensors.

An example of the latter issue would be the infinite number of different sensations caused by an object on a visual system as it is translated or rotated in space. Each different pose (i.e. position and orientation) of the object relative to the observer will sensitise the photoreceptor grid of the imaging sensor in a different way, so an impossible number of projections of the object on the image-plane would have to be used as features for classification, if no preprocessing to achieve pose-invariance perception was to be performed.

This implies that, in general, Bayesian classifiers in their simplest form are seldom used in modern probabilistic frameworks of artificial perception. Modularity must step in, through the introduction of intermediate variables – this will be further investigated in Chapter 4.

2.7 Final Remarks and Further Reading

A lot more can be said regarding all aspects of spatial referencing. The interested reader should be, for example, aware of the work of Klatzky [25], of Byrne and Becker [9] and from groups such as those of Meilinger, Bülthoff and Berthoz et al. [13] regarding the significance of and relations between frames of reference and points of view.

As for spatial mapping, we have just skimmed the surface of this subject in this chapter. In fact, as you might recall we have analysed mapping in its abstract form as a perceptual representation of the environment. Spatial maps should be associated, however, not only with the sensation that produces them (the "input" perspective), but also with the decisions and actions that will be performed on the objects that populate them (the "output" perspective) – we will therefore recurrently return to this subject in the following chapters. However, as a starting reference in this matter, we refer the reader to the original seminal work of Moravec and Elfes [40; 38; 36] on occupancy and inference grids, and important recent developments such as the work by Wurm et al. [6], who use an octotree-based grid. In any case, we will be revisiting the occupancy map with an in-depth look on its generative model in the following chapter.

Finally, Colas, Diard, and Bessière [4] an excellent, concise reference article presenting, amongst others, common Bayesian models in the context of sensing and perception.

References

1. Ferreira, J.F., Lobo, J., Bessiére, P., Castelo-Branco, M., Dias, J.: A Bayesian Framework for Active Artificial Perception. IEEE Transactions on Systems, Man, and Cybernetics, Part B: Cybernetics 43(2), 699–711 (2013) ISSN 1083-4419, doi:10.1109/TSMCB.2012.2214477 44
2. Portugal, D., Rocha, R.P.: Topological Information from Grid Maps for Robot Navigation. In: Proceedings of the 4th International Conference on Agents and Artificial Intelligence (ICAART 2012), Vilamoura, Portugal, pp. 137–143 (2012) 46
3. Ranganathan, A., Dellaer, F.: Online Probabilistic Topological Mapping. International Journal of Robotics Research 30(6), 755–771 (2011) 46, 47
4. Colas, F., Diard, J., Bessiére, P.: Common Bayesian Models For Common Cognitive Issues. Acta Biotheoretica 58(2-3), 191–216 (2010) 50, 67
5. Faria, D.R., Martins, R., Lobo, J., Dias, J.: Probabilistic Representation of 3D Object Shape by In-Hand Exploration. In: Proceedings of The 2010 IEEE/RSJ International Conference on Intelligent Robots and Systems, IROS 2010, Taipei, Taiwan (2010) XXI, 56, 57, 58
6. Wurm, K.M., Hornung, A., Bennewitz, M., Stachniss, C., Burgard, W.: OctoMap: A Probabilistic, Flexible, and Compact 3D Map Representation for Robotic Systems. In: Proceedings of the ICRA 2010 Workshop on Best Practice in 3D Perception and Modeling for Mobile Manipulation (2010) 66
7. Ferreira, J.F., Pinho, C., Dias, J.: Implementation and Calibration of a Bayesian Binaural System for 3D Localisation. In: 2008 IEEE International Conference on Robotics and Biomimetics (ROBIO 2008), Bangkok, Thailand (2009) 59, 61, 62, 63
8. Burgess, N.: Spatial Cognition and the Brain. Annals of the New York Academy of Sciences 1124, 77–97 (2008) 38, 39
9. Byrne, P., Becker, S.: A principle for learning egocentric-allocentric transformation. Neural Computation 20(3), 709–737 (2008) 39, 66
10. Ferreira, J.F., Bessiére, P., Mekhnacha, K., Lobo, J., Dias, J., Laugier, C.: Bayesian Models for Multimodal Perception of 3D Structure and Motion. In: International Conference on Cognitive Systems (CogSys 2008), pp. 103–108. University of Karlsruhe, Karlsruhe (2008) 44
11. Wiener, J.M., Meilinger, T., Berthoz, A.: The integration of spatial information across different perspectives. In: Proceedings of the 30th Annual Conference of the Cognitive Science Society (CogSci 2008), pp. 2031–2036 (2008) 39
12. Doya, K., Ishii, S., Pouget, A., Rao, R.P.N. (eds.): Bayesian Brain — Probabilistic Approaches to Neural Coding. MIT Press (2007) ISBN 978-0-262-04238-3 51
13. Meilinger, T., Riecke, B.E., Bülthoff, H.H.: Orientation Specificity in Long-Term-Memory for Environmental Spaces. In: Proceedings of the 29th Annual Conference of the Cognitive Science Society (CogSci 2007), pp. 479–484 (2007) 39, 66

14. Tapus, A., Battaglia, F., Siegwart, R.: The Hippocampal Place Cells and Fingerprints of Places: Spatial Representation Animals, Animats and Robots. In: Proceedings of the 9th International Conference on Intelligent Autonomous Systems, IAS-9 (2006) 49
15. Vasudevan, S., Nguyen, V., Siegwart, R.: Towards a Cognitive Probabilistic Representation of Space for Mobile Robots. In: Proceedings of the IEEE International Coference on Information Acquisition (ICIA), Shandong, China (2006) 49
16. Taddeo, M., Floridi, L.: Solving the symbol grounding problem: a critical review of fifteen years of research. Journal of Experimental and Theoretical Artificial Intelligence 17(4), 419–445 (2005) 37
17. Faller, C., Merimaa, J.: Source localization in complex listening situations: Selection of binaural cues based on interaural coherence. Journal of the Acoustical Society of America 116(5), 3075–3089 (2004), doi:10.1121/1.1791872 61, 64
18. Kapralos, B., Jenkin, M.R.M., Milios, E.: Auditory Perception and Spatial (3D) Auditory Systems. Technical Report CS-2003-07, York University (2003) 60
19. Tomatis, N., Nourbakhsh, I., Siegwart, R.: Hybrid simultaneous localization and map building: a natural integration of topological and metric. Robotics and Autonomous Systems 44(1), 3–14 (2003) 48, 49
20. Cohen, Y.E., Anderson, R.E.: A Common Reference Frame for Movement Plans in the Posterior Parietal Cortex. Nature Reviews Neuroscience 3, 553–562 (2002) 39
21. Gallistel, C.R.: Language and spatial frames of reference in mind and brain. TRENDS in Cognitive Sciences 6(8), 321–322 (2002) 38
22. King, A.J., Schnupp, J.W., Doubell, T.P.: The shape of ears to come: dynamic coding of auditory space. TRENDS in Cognitive Sciences 5(6), 261–270 (2001) 60
23. Shinn-Cunningham, B.G., Santarelli, S., Kopco, N.: Tori of confusion: Binaural localization cues for sources within reach of a listener. Journal of the Acoustical Society of America 107(3), 1627–1636 (2000) 60, 61
24. Colby, C.L.: Action-Oriented Spatial Reference Frames in Cortex. Neuron 20, 15–24 (1998) Review 39
25. Klatzky, R.L.: Allocentric and egocentric spatial representations: Definitions, distinctions, and interconnections. In: Freksa, C., Habel, C., Wender, K.F. (eds.) Spatial Cognition 1998. LNCS (LNAI), vol. 1404, p. 1. Springer, Heidelberg (1998) 38, 39, 66
26. McIntyre, J., Stratta, F., Lacquaniti, F.: Short-Term Memory for Reaching to Visual Targets: Psychophysical Evidence for Body-Centered Reference Frames. Journal of Neuroscience 18(20), 8423–8435 (1998) 44
27. Murphy, K.J., Carey, D.P., Goodale, M.A.: The Perception of Spatial Relations in a Patient with Visual Form Agnosia. Cognitive Neuropsyshology 15(6/7/8), 705–722 (1998) 44
28. Thrun, S.: Learning metric-topological maps for indoor mobile robot navigation. Artificial Intelligence 99(1), 21–71 (1998) 47, 49
29. Cutting, J.E., Vishton, P.M.: Perceiving layout and knowing distances: The integration, relative potency, and contextual use of different information about depth. In: Epstein, W., Rogers, S. (eds.) Handbook of Perception and Cognition, vol. 5, Academic Press (1995) Perception of space and motion 44

30. Gordon, J., Ghilardi, M.F., Ghez, C.: Accuracy of planar reaching movements. I. Independence of direction and extent variability. Experimental Brain Research 99(1), 97–111 (1994) 44
31. Kuipers, B., Byun, Y.T.: A robot exploration and mapping strategy based on a semantic hierarchy of spatial representations. Robotics and Autonomous Systems 8, 47–63 (1991) 45, 48
32. Leonard, J.J., Durrant-Whyte, H.F.: Mobile Robot Localization by Tracking Geometric Beacons. IEEE Transactions on Robotics and Automation 7(3), 376–382 (1991) 41
33. Viemeister, N.F., Wakefield, G.H.: Temporal integration and multiple looks. The Journal of the Acoustical Society of America 90(2), 858–865 (1991) 61
34. Gallistel, C.R.: The organization of learning. Learning, development, and conceptual change. MIT Press, Cambridge (1990) 38
35. Harnad, S.: The symbol grounding problem. Physica D 42, 335–346 (1990) 37
36. Elfes, A.: Using occupancy grids for mobile robot perception and navigation. IEEE Computer 22(6), 46–57 (1989) 42, 66
37. Oppenheim, A.V., Schafer, R.: Discrete-Time Signal Processing (1989) 61
38. Moravec, H.P.: Sensor fusion in certainty grids for mobile robots. AI Magazine 9(2), 61–74 (1988) 42, 66
39. Zurek, P.M.: The precedence effect. In: Yost, W., Gourevitch, G. (eds.) Directional Hearing. Springer (1987) 61
40. Moravec, H., Elfes, A.: High resolution maps from wide angle sonar. In: IEEE International Conference on Robotics and Automation (1985) 42, 66

3

Bayesian Programming and Modelling

So far as the theories of mathematics are about reality, they are not certain; so far as they are certain, they are not about reality.

Sidelights on Relativity (Geometry and Experience), *Albert Einstein (1923)*

Mathematics is the art of giving the same name to different things.

Science and method, *Henri Poincaré (1914)*

A problem well stated is a problem half solved.

Charles Kettering (1876-1958)

3.1 Introduction

A vast amount of different formalisms exist for the construction of probabilistic models (Fig. 3.1):

- General formalisms, which allow the construction of more encompassing and potentially more complete models.
- Specific formalisms, which yield simpler or more intuitive formulations, thus allowing for easier or more efficient computation.

An essential step to adopting probabilistic approaches to robotic perception is to become aware of the range of available modelling formalisms and also of the most important inference techniques that support model implementation.

We have briefly introduced Bayesian networks in Chapter 1; we will cover this subject in more detail on the following section, and present several supporting examples. Next, we will present the notion of probabilistic loops, for which we will introduce the concepts behind the respective formalisms. Subsequently, we will present the ultimate generalisation for Bayesian modelling – the Bayesian programming formalism – and discuss its advantages comparing to graphical models. Finally, we will offer an overview of Bayesian inference techniques and very brief list of useful implementation tools available to the modeller.

J.F. Ferreira and J. Dias: *Probabilistic Approaches for Robotic Perception*,

Springer Tracts in Advanced Robotics 91,
DOI: 10.1007/978-3-319-02006-8_3,

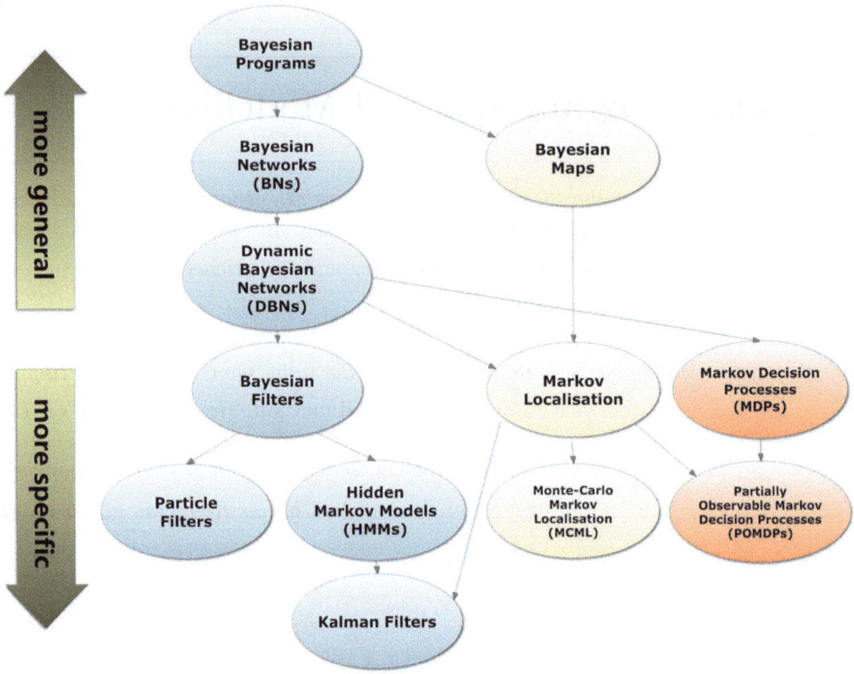

Fig. 3.1. Taxonomy of Bayesian formalisms for probabilistic model construction (adapted from [7]). As shown in the diagram, these formalisms range from general to specific, and the arrows show how specific formalisms are derived from and related to general formalisms. For example, a dynamic Bayesian network is capable of formalising the same model as a Bayesian filter (the arrow flows from the former to the latter), but not all Bayesian networks can be represented as dynamic Bayesian networks (the arrow flows from the latter to the former). Formalisms on the far left and centre left lanes have generic applications: they will be introduced in this chapter. Formalisms in the centre right lane are used for mapping and localisation applications, while formalisms in far right lane are used for decision processes, and both will be presented in Chapter 5.

3.2 Bayesian Formalisms for Probabilistic Model Construction

3.2.1 Bayesian Networks Revisited and the Plate Notation

Also called belief networks, Bayesian networks (BNs), as we have already seen in Chapter 1, are graphical models that represent a set of random variables and their conditional dependencies via directed acyclic graphs (DAG – directed graphs with no loops formed by directed cycles). Bayesian networks generally represent causal relationships through the directed edges, but there have been exceptions where they represent dependences, i.e., the inverse direction.

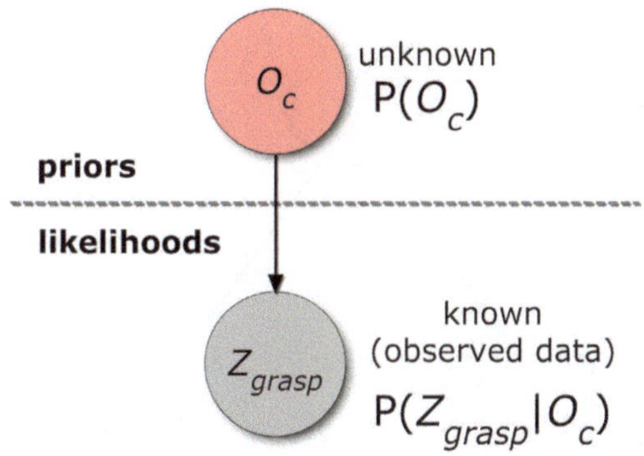

Fig. 3.2. Bayesian network for the occupancy grid model used by Faria et al. [4] for object representation using in-hand exploration. Prior, posterior and respective distributions have optionally been labelled for better illustration of the graphical representation; in general, they are omitted, since they are implicit. Note that the variables and distributions themselves are not specified; they are only defined in terms of their notation and conditional dependence. However, the model will not be complete until these distributions are specified, the way they will be instantiated is defined (i.e. their parameters), and the random variables that support the model are fully described (i.e. their significance and measurable space).

Let us now see an example of the Bayesian network formalism applied to the decomposition of the joint distribution in which the sensor model by Faria et al. [4] is used, presented in the previous chapter on Example 2.2.

Example 3.1. **Bayesian network for the in-hand exploration of objects**

Faria, Martins, Lobo, and Dias [4] applied their sensor model to a common occupancy grid framework, whose decomposition equation for its joint distribution is given by

$$P(Z_{grasp} \wedge O_c) = P(O_c)P(Z_{grasp} \mid O_c).$$

This is a very simple decomposition equation and, as such, results in a very simple graphical representation as a Bayesian network, following the basic notation rules described on Chapter 1. This representation is given on Fig. 3.2.

Buntine [31] introduced a useful add-on for Bayesian networks: the *plate notation*. The plate notation is a method of representing variables that repeat in a graphical model. Instead of drawing each repeated variable individually, a plate or rectangle can be used to group variables into a subgraph.

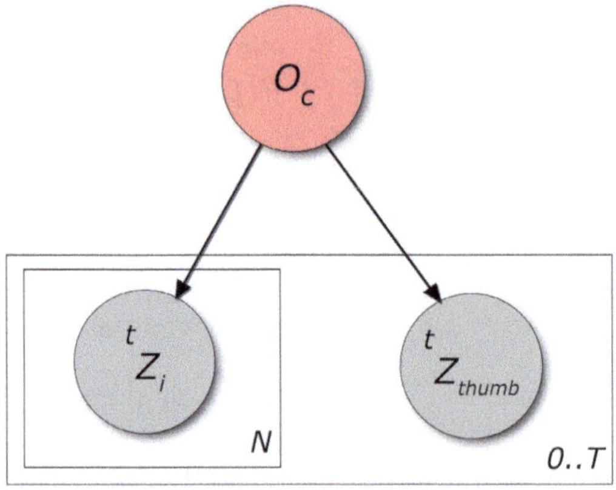

Fig. 3.3. Contribution of the sensor on each finger through time made explicit using the Bayesian network formalism with plate notation applied to the example of in-hand exploration of objects by Faria et al. [4].

The plate notation relies on the following assumptions:

- The subgraph is duplicated as many times as the associated repetition number.
- The variables in the subgraph are indexed up to the repetition number.
- The respective distributions appear in the joint distribution as an *indexed product of the sequence of variables.*
- The links that cross a plate boundary are replicated for each subgraph repetition.

The following example illustrates the Bayesian network formalism being used in combination with the plate notation.

Example 3.2. **Bayesian network with plate notation for the in-hand exploration of objects**

Expanding on the model of Example 3.1, the contribution of the sensor on each finger through time can be made explicit on the decomposition equation as follows

$$P({}^0Z_{thumb} \wedge \ldots \wedge {}^TZ_{thumb} \wedge {}^0Z_0 \wedge \ldots \wedge {}^TZ_0 \wedge \ldots \wedge {}^0Z_N \wedge \ldots \wedge {}^TZ_N \wedge O_c) =$$
$$P(O_c)\prod_{t=0}^{T} P({}^tZ_{thumb} \mid O_c)\prod_{i=1}^{N} P({}^tZ_i \mid O_c),$$

with T and $N = 4$ representing the current time instant and the remaining four fingers of the hand, respectively.

Its representation as a Bayesian network, using the plate notation, is given on Fig. 3.3.

3.2.2 Probabilistic Loops: Dynamic Bayesian Networks and Bayesian Filtering

It is almost inconceivable to think of cognitive systems, and in particular perceptual systems, that simply take data as an input and, in a single, unidirectional stream of computation, directly produce an output. Most of these systems rely on mechanisms of information feedback, that allow the framework to reassess its current cognitive state – a never ending set of examples can be found in artificial perception systems, and also in natural cognitive systems. This is analogous to closed-loop systems in control theory – consequently, applying this notion to probabilistic modelling, we are in the presence, in these cases, of *probabilistic loops*. Note, however, that each random variable in a model may only appear once in the left side of the conditioning bar, which in turn means that direct probabilistic loops at first glance are impossible. This fact is circumvented by replicating those variables (nodes in a BN) and explicitly assuming a set of circumstances that differentiate the replicated instances (e.g. time instants), thus creating an indirect loop – you will see this in practice in the remainder of the section.

When the independence assumptions do not vary over time, this class of models is called *dynamic Bayesian networks* (DBN). Dynamic Bayesian networks are a specific type of Bayesian networks that represent sequences of variables – these sequences are usually time-series, but can also be sequences of symbolic entities. Therefore, the following definitions, while generally applicable, assume this temporal dependency in their description.

Let ${}^{0:T}O$ denote a time series of observation variables from time 0 to T, i.e., ${}^{0:T}O \equiv {}^{0}O \wedge {}^{1}O \wedge \ldots \wedge {}^{T}O$. Let ${}^{0:T}S$ represent the state variables[1] over the same time period. The general model class is given by

$$P({}^{0:T}O \wedge {}^{0:T}S) = P({}^{0}S \wedge {}^{0}O) \prod_{t=1}^{T} P({}^{t}S \wedge \ldots \wedge {}^{0}S \wedge {}^{t}O). \qquad (3.1)$$

As mentioned above, DBNs are mainly used for recursive Bayesian estimation over a time period, which means that they are implemented using *memory*. A useful simplification for (3.1) is to make the Markov assumption (Chapter 1); if we assume a first-order Markov model, we obtain

$$P({}^{0:T}O \wedge {}^{0:T}S) = P({}^{0}S \wedge {}^{0}O) \prod_{t=1}^{T} P({}^{t}S \wedge {}^{t-1}S \wedge {}^{t}O), \qquad (3.2)$$

$$P({}^{0:T}O \wedge {}^{0:T}S) = P({}^{0}S \wedge {}^{0}O) \prod_{t=1}^{T} P({}^{t}S \mid {}^{t-1}S) P({}^{t}O \mid {}^{t}S). \qquad (3.3)$$

[1] Referring to the state of the surrounding environment, or the state of an observed object, or the phyiscal state or the state of mind of a robot's interlocutor, etc.

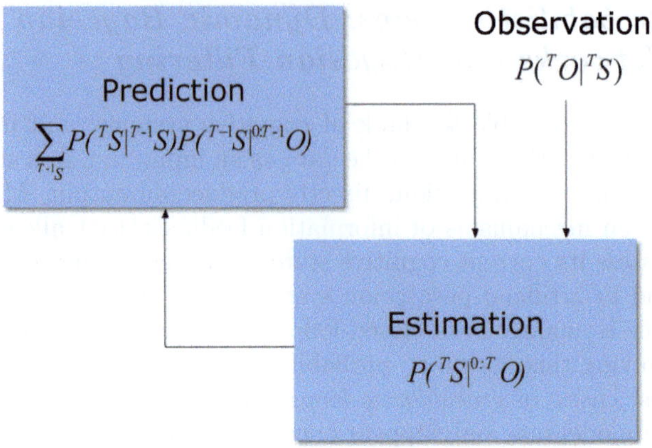

Fig. 3.4. The Bayesian filter loop. As can be seen in this graphical representation, the observation, added to a prediction process given by the dynamic model, gives rise to an estimate, which in turn fuels the prediction process for the next inference step. Note that the word "filter" in this context, although historically tied to the filtering concept, has evolved to refer to the general time-invariant probabilistic loop model.

In this case, the space for storing ^{t-1}S is taken as being enough memory to accommodate the system.

Equation (3.3) represents a further assumption – the *stationarity hypothesis* – which states that, for the entire time sequence, the dependency structure of the variables remains the same. In this equation, $P(^tS \mid {}^{t-1}S)$ is usually called the dynamic or *transition model*, while $P(^tO|^tS)$ is called the *observation model*.

When both the transition and observation models have the same form for all time steps, t, the local model, $P(^tO \wedge {}^tS \wedge {}^{t-1}S) = P(^tS \mid {}^{t-1}S)P(^tO \mid {}^tS)P(^{t-1}S)$, is said to be *time-invariant* or *homogeneous*.

State estimation, which corresponds to answering the question, $P(^TS \mid {}^{0:T}O)$, is solved through

$$\underbrace{P(^TS \mid {}^{0:T}O)}_{\textbf{estimation}} \propto \overbrace{P(^TO \mid {}^TS)}^{\text{observation model}} \underbrace{\sum_{^{T-1}S} \overbrace{P(^TS \mid {}^{T-1}S)}^{\text{transition model}} \overbrace{P(^{T-1}S \mid {}^{0:T-1}O)}^{\text{prior}}}_{\textbf{prediction}} . \tag{3.4}$$

Therefore, the computation of state estimation at time T is performed recursively, based on $P(^{T-1}S \mid {}^{0:T-1}O)$, the answer to the state estimation question on the preceding inference step (i.e the *prior* on the state).

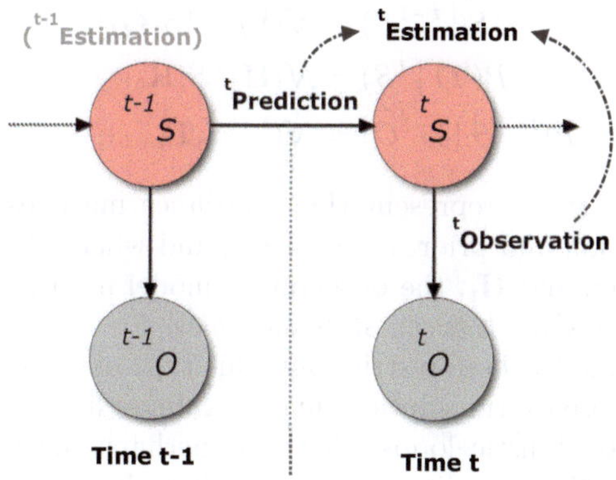

Fig. 3.5. Generic dynamic Bayesian network for the Hidden Markov model. This is the simplest instantiation of a DBN, and it represents one of the few cases for which exact inference is feasible. Note that, while the state space is, by definition, discrete, the observation space may be either discrete (and in that case, generally categorical), or continuous (and in that case, often normally distributed).

In a nutshell, usually the state is reevaluated each time a new observation is acquired, using the past state estimate as well as the prediction and observation models, although in some cases (for example, to keep a regular estimation rate), the state might be at times reevaluated using prediction alone. In other words, *state estimation occurs in a loop over time* – see Fig. 3.4. This same model can be used to *predict future states* ($P({}^{t+k}S \mid {}^{0:t}O), k > 0$), or to refine – i.e. *filter* – a past estimate given observations that occur later in time ($P({}^{t-k}S \mid {}^{0:t}O), k > 0$) [3]. The larger the value of k, the more computationally expensive these mechanisms become, as their respective inference processes require summations over $k-1$ variables. The filtering notion gave rise to the general case denomination *Bayesian filters* for this type of models.

When the state space can be assumed to be discrete, *hidden Markov models* (HMM) can be applied – see Fig. 3.5. Another common technique, used for approximating the inference required for estimation when the state space is assumed continuous in highly nonlinear or computationally expensive cases, is to model the state probability distribution using *particle filters*. Particle filters are so named because they allow for approximate "filtering" (in the general, Bayesian filtering sense) by generating and using a set of "particles" – differently-weighted samples that approximate the posterior distribution.

Exact, closed-form solutions to the recursive computation exist in a restrict set of cases, when the state space is continuous. In particular, the *Kalman filter* is an optimal solution when *all models are assumed to follow normal distributions* described as

$$P({}^tS \mid {}^{t-1}S) = \mathcal{N}(\mathbf{F}_t\, {}^{t-1}S, \mathbf{Q}_t), \tag{3.5a}$$

$$P({}^tO \mid {}^tS) = \mathcal{N}(\mathbf{H}_t\, {}^tS, \mathbf{R}_t), \tag{3.5b}$$

$$P({}^{t-1}S \mid {}^{t-1}O) = \mathcal{N}({}^{t-1}\hat{S}, \mathbf{P}_{t-1}), \tag{3.5c}$$

where $\mathbf{Q}_t$, $\mathbf{R}_t$ and $\mathbf{P}_{t-1}$, represent the covariance matrices of the process noise, the observation and prior, respectively, and where $\mathbf{F}_t$, the state transition model matrix, and $\mathbf{H}_t$, the observation model matrix, are *assumed to define linear relationships between state variables.*

The fact that, due to these restrictions, this type of model even so represents an optimal closed-form solution, makes Kalman filters maybe the most widely used of probabilistic loops. However, perhaps the most limiting of these restrictions is the linearity assumption, since, in general, most systems are highly nonlinear. This nonlinearity may be associated either with the process model or with the observation model or with both. Extensions and generalizations to this method have thus been developed, such as the *extended Kalman filter* (EKF) and the *unscented Kalman filter* (UKF); however, none of these is an optimal solution. In fact, particle filters, with sufficient samples, approach the Bayesian optimal estimate, so they can be made more accurate than either the EKF or UKF.

Finally, a last remark about observations in a probabilistic loop. Observations may result either from *hard evidence* or *soft evidence*. Hard evidence allows specifying the instantiation of an observation variable directly. Soft evidence, on the other hand, allows the description of probability distributions for the observation variable, but not the actual instantiation of the observation variable directly. So, in simple terms, hard evidence on an observation means that the actual value of the corresponding variable is made available to the model, while soft evidence on an observation means that a set of plausibility values concerning each value of the observation variable is made available to the model. In any case, any of these types of evidence allow for Bayesian inference.

3.2.3 The Generalisation: Bayesian Programming

The *Bayesian program* (BP), as first defined by Lebeltel [28] and later consolidated by Bessière, Laugier, and Siegwart [7], is a generic formalism for building probabilistic models and for solving decision and inference problems on these models. This formalism was created to supersede, restate and compare numerous classical probabilistic models such as Bayesian networks, Dynamic Bayesian networks, Bayesian Filters, Hidden Markov Models, Kalman Filters, Particle Filters, Mixture Models, or Maximum Entropy Models, as shown on Fig. 3.1.

A Bayesian program consists of two parts (Fig. 3.6):

- a *description* which is the probabilistic model of the studied phenomenon or programmed behaviour;

- Program
  - Description
    - Specification
      - Relevant variables:
        $X_1, X_2, \ldots, X_N$
      - Decomposition:
        $P(X_1 \wedge X_2 \wedge \ldots \wedge X_N \mid \pi) =$
        $P(L_0 \mid \pi)P(L_1 \mid L_0 \wedge \pi) \ldots P(L_K \mid L_{K-1} \wedge L_{K-2} \ldots L_0 \wedge \pi) =$
        $P(L_0 \mid \pi)P(L_1 \mid R_1 \wedge \pi) \ldots P(L_K \mid R_K \wedge \pi)$
      - Parametric forms:
        $P(L_i \mid R_i \wedge \pi), \forall i \in 0 \ldots K$: distribution of type $f_\mu(L_i)$,
        where μ is a vector of parameters that may depend either on R_i, on experimental data, or both.
    - Identification:
      Method for estimating free parameters of $f_\mu(L_i)$.
  - Question:
    $P(Search \mid known \wedge \pi)$, where $Search$ and $known$ (with $known$ being a particular instantiation of $Known$) are conjunctions of subsets of relevant variables.

Fig. 3.6. Generic Bayesian program. See main text for the definition of auxiliary variables $L_0, \ldots, L_K$ and their corresponding counterparts $R_0, \ldots, R_K$.

- a *question* that specifies an inference problem to be solved using this model.

The description itself contains two subparts:

- a *specification* section that formalises the knowledge of the programmer;
- an *identification* section, in which the model's free parameters are predefined by the programmer, or the procedure for estimating the model's free parameters from experimental data is specified.

In the following text, each of the constituents of a generic Bayesian program will be explained in greater detail, based on what is presented on Bessière et al. [7].

Description

As already defined, the description is the probabilistic model of the studied phenomenon or programmed behaviour. All the knowledge available about this phenomenon or behaviour is encoded in the *joint probability distribution* on the *relevant variables* (see Fig. 3.6).

Unfortunately, this joint distribution is generally too complex to use as is. The first purpose of the description is to give an effective method of computing the joint distribution in a tractable fashion (specification). The second purpose is to specify the learning methods for identifying values of the free parameters from the observed data (identification).

Specification

The programmer's knowledge is specified in a sequence of three steps:

1. *Define the set of relevant variables* $\{X_1, X_2, \ldots, X_N\}$ on which the joint distribution is defined.
2. *Decompose the joint distribution* to obtain a tractable way to compute it. The only rule that must be obeyed to attain a valid probabilistic expression is that each variable must appear only once on the left side of the conditioning bar (sometimes called the *chain-rule of decomposition*). This is formally expressed as follows. Given a partition of $\{X_1, X_2, \ldots, X_N\}$ into K subsets, we define K variables $L_0, \ldots, L_K$, each corresponding to one of these subsets. Each variable L_i is consequently obtained as the conjunction of the variables composing each subset i.
   The recursive application of the conjunction rule (Chapter 1) leads to the exact mathematical expression on L_i presented in Fig. 3.6.
   On the other hand, *conditional independence* hypotheses then allow for further simplifications. Such a hypothesis can be defined for variable L_i by picking a subset of variables X_j among the variables appearing in the conjunction formed by $L_{i-1} \ldots L_0$ by denoting the latter by R_i and rewriting the joint distribution decomposition as shown on Fig. 3.6.
3. *Define the parametric forms* that give an explicit means to compute each distribution $P(L_i \mid R_i \wedge \pi)$ appearing in the decomposition. This is achieved by associating each distribution $P(L_i \mid R_i \wedge \pi)$ with a function $f_\mu(L_i)$ — μ denotes the set of parameters that define the distribution — or a question to another Bayesian program.

Identification

The role of the identification phase is to assign values to free parameters within the set μ, either through direct assignment or through the estimation of these parameters using *Bayesian learning* with experimental data.

Question

Given a particular description on a BP, a question is obtained by partitioning the set of relevant variables into three sets: the *searched variables* (the conjunction of which is denoted by $Search$), the *known variables* (the conjunction of which is denoted by $Known$) and the *free variables* (the conjunction of which is denoted by $Free$).

For a given value of the variable $Known$ (denoted by $known$), a question is defined as $P(Search \mid known \wedge \pi)$, as shown on Fig. 3.6.

3.2.4 Bayesian Programming vs. Bayesian Networks

At first, the programming syntax presented in the previous section might seem less convenient than the graphical interface of standard Bayesian network formalisms.

Bessière, Laugier, and Siegwart [7] argue that "the absence of an evident man-machine interface was not a negligence but a choice":

- Graphical representations impose supplementary constraints that result, neither from the rules of probability, nor from the logic of the problem. For instance, the Bayesian approach allows to specify decompositions including distributions with two or more variables on the left part of the conditioning mark (e.g $P(X \wedge Y \mid \pi)$). This is not possible using the Bayesian network formalism without introducing an intermediate variable. This also means that, while all Bayesian network models may be restated as Bayesian program, the opposite is not always true.
- The algebraic notation used in BP is very convenient to express iteration or recursion. This greatly simplifies the specification of models that include submodels duplicated several times such as, for instance, hierarchical versions of Bayesian filters or Hidden Markov Models (Chapter 4).
- Bayesian programs, as implied by the name, have been devised so as to be easily translated into computer programming constructs falling into the category of the *declarative programming paradigm.* This opens the possibility of thinking in terms of conditional constructs used in computer programming, such as "if-then-else" and, "subroutine calls" (see Chapter 4).

Of course, both formalisms offer different advantages, and since most models are translatable from one to the other, the reader should use them as fits the opportunity presented by the problem at hand.

3.3 Bayesian Inference Techniques and Model Implementation

We can now define Bayesian inference, in very concrete, formal terms, as the process of answering the question $P(Search \mid Known)$ (i.e. determining the posterior), by computing

$$P(Search \mid Known) \propto \sum_{Free} P(Search \wedge Known \wedge Free). \tag{3.6}$$

Inference can be either exact or approximate; both flavours will be discussed in the following subsections.

3.3.1 Exact Inference

When it is possible to compute the posterior using a closed-form solution, inference is said to be exact. *Exact inference* is only feasible for a very limited set of cases, namely:

- when *all free variables are discrete* (e.g. hidden Markov models);

- when *all distributions in the decomposition equation are linear and normal* (e.g. Kalman filters).

Exact inference algorithms can be separated into those that only work on DAG models and general solutions. The former, called *variable elimination algorithms*, work by exploiting the chain-rule of decomposition (see section 3.2.3) to "push sums inside products" and marginalise out the unwanted free variables. General exact inference algorithms, on the other hand, are defined in terms of message passing on a tree – see, for example, Pearl's algorithm, on which the original graph is converted to a junction tree using cutset conditioning [35].

3.3.2 Approximate Inference

Even in situations where exact inference is mathematically possible, it might not be computationally tractable. In fact, there are two reasons why *approximate inference* might be necessary: the computation required for exact inference might be too lengthy, or there might be no analytic solution whatsoever for a given model.

Three techniques have been widely used in such cases:

- **Sampling methods.** Also called *Monte Carlo methods*, the simplest kind is *importance sampling*, where random samples x are drawn from the prior $P(X)$ and then the samples are weighted by their likelihood, $P(y \mid x)$, where y is the evidence. The dominant method within these methods would be the Monte Carlo Markov Chain (MCMC), usually more efficient for higher dimensions. The particle filter, briefly discussed in section 3.2.2, is a *sequential Monte Carlo method* (SMC).
- **Variational methods.** In their simple form (the *mean-field approximation*), all random variables are decoupled by introducing the so-called *variational parameters* for each, which will be iteratively updated so as to minimise the Kullback-Leibler divergence (Chapter 1) between the approximate and the true probability distributions. The approximation consequently produces a lower bound on the likelihood. These methods have been applied to approximate Bayesian inference by means of a technique called Variational Bayes [22].
- **Belief propagation** (BP). This method basically applies the message passing algorithm to the original graph instead of a tree. This technique, originally used for exact inference only, was later extended to perform approximate Bayesian inference using a method called Expectation Propagation [23].

3.3.3 Software for Model Implementation

There are numerous software packages for performing Bayesian inference, and several extensive lists have been compiled on this subject – see, for example,

[24] and [2]. Despite not being within the scope of this book, we thought it would be interesting for the reader to be made aware of a few packages that offer an application programming interface (API) specifically for *model implementation.*

The most suitable and/or well-known APIs for model implementation in the authors view would be the following:

- The **Bayes Net Toolbox for MATLAB** (BNT), by Murphy [24], an open-source MATLAB package for DAGs. The BNT supports many kinds of nodes (probability distributions), exact and approximate inference, parameter and structure learning, and static and dynamic models.
- The **BUGS (Bayesian inference Using Gibbs Sampling)** software package [6]. The user specifies a statistical model, of (almost) arbitrary complexity, by simply stating the relationships between related variables. The software includes an "expert system", which determines an appropriate MCMC scheme (based on the Gibbs sampler) for analysing the specified model. The user then controls the execution of the scheme and is free to choose from a wide range of output types.
- The **ProBT® library**, by ProBAYES [13]. ProBT® aims at providing a programming tool that facilitates the creation of Bayesian models and their reusability using the Bayesian programming formalism presented on section 3.2.3. This property allows the design of advanced features such as submodel reuse, learning, and distributed inference. It consists of two layers: (1) the ProBT® Engine, the core of the library, a set of high-performance inference algorithm modules developed in C++ language, which runs on the most common operating systems, such Linux/Unix, Windows, and MacOS X; (2) the ProBT API, an application programming interface available in C++ and Python for accessing the ProBT® Engine functions.

Although all of these packages have their own merits, the ProBT® library presents the advantage of offering a very powerful way of defining conditional probability distributions, so that the parameters of these distributions can be directly related to the variables on the right of the conditioning bar, in a very user-friendly fashion, through external functions on these parameters. On the other hand, as described on section 3.2.4, there are a few important advantages on using the Bayesian programming formalism as opposed to Bayesian networks.

3.4 Bayesian Modelling for Robotic Perception

In the following text, we will go through a few exemplary Bayesian models for robotic perception using the techniques learned in this chapter. We will start by returning to the occupancy grid in order to fully define its generic model. Next, we will analyse two published worked out examples.

3.4.1 The Occupancy Grid Revisited

The traditional occupancy grid framework, as mentioned in the previous chapter, is a Markov random field, which means that the occupancy state of each cell c, denoted as the binary random variable O_c, is independent from the states of all other cells in the grid. Due to this assumption, the generative model for each cell is very simply stated at a particular time instant t as

$$P(O_c \wedge Z_1 \wedge \ldots \wedge Z_N) = P(O_c) \prod_{i=1}^{N} P(Z_i \mid O_c), \tag{3.7}$$

where $P(Z_i \mid O_c)$ is the generic direct sensor model on occupancy for each of the N types of sensations Z_i available for processing at that time instant. Moreover, since temporally the occupancy grid is traditionally a first-order Markov process, the prior $P(O_c)$ in this equation is taken as the posterior $P(O_c \mid Z_1 \wedge \ldots \wedge Z_N)$ obtained in the previous time instant, $t-1$. It constitutes, therefore, a very simple Bayesian filter.

Applying Bayes' rule on equation (3.7), the following posteriors on the two possible states of occupancy of a specific cell c, with $[O_c = 1]$ meaning that the cell is occupied and $[O_c = 0]$ meaning that the cell is empty,

$$P([O_c = 1] \mid Z_1 \wedge \ldots \wedge Z_N) = \frac{P([O_c = 1]) \prod_{i=1}^{N} P(Z_i \mid [O_c = 1])}{\sum_{O_c} P(O_c) \prod_{i=1}^{N} P(Z_i \mid O_c)}, \tag{3.8a}$$

and

$$P([O_c = 0] \mid Z_1 \wedge \ldots \wedge Z_N) = \frac{P([O_c = 0]) \prod_{i=1}^{N} P(Z_i \mid [O_c = 0])}{\sum_{O_c} P(O_c) \prod_{i=1}^{N} P(Z_i \mid O_c)}. \tag{3.8b}$$

Let us now define the *odds* of a generic binary random variable variable A as

$$odds(A) = \frac{P([A = 1])}{P([A = 0])}. \tag{3.9}$$

Examining closely the odds of the occupancy state, it becomes clear that $odds(O_c) \in [0, +\infty)$. Since $P([O_c = 0]) + P([O_c = 1]) = 1$, this means that the odds compactly represents the occupancy state, with the added benefit of an expanded range which allows for a more precise representation as a floating point number by the robot's computational resources in practical applications.

Next, let us define the *log-odds* of a generic binary random variable A as

$$\ln odds(A) = \ln \frac{P([A=1])}{P([A=0])}. \tag{3.10}$$

Following up with the previous reasoning, $\ln odds(O_c) \in (-\infty, +\infty)$, which means an even greater range; note also that $\ln odds(O_c) = 0$ means maximum entropy/uncertainty, in other words, $P([O_c = 1]) = P([O_c = 0]) = .5$. Moreover, log-probability arithmetic, as mentioned already in Chapter 1, is, in practice, much less computationally expensive than using regular probabilities, since products are transformed into sums, which demand much less computational power.

In fact, considering

$$\ln odds(O_c \mid Z_1 \wedge \ldots \wedge Z_N) = \ln \frac{P([O_c = 1] \mid Z_1 \wedge \ldots \wedge Z_N)}{P([O_c = 0] \mid Z_1 \wedge \ldots \wedge Z_N)}$$

and

$$\ln \lambda(Z_i \mid O_c) = \ln \frac{P(Z_i \mid [O_c = 1])}{P(Z_i \mid [O_c = 0])},$$

the inference equations (3.7) are more compactly rewritten as a single, simple and extremely efficient update equation

$$\ln odds(O_c \mid Z_1 \wedge \ldots \wedge Z_N) = \ln odds(O_c) + \sum_{i=1}^{N} \ln \lambda(Z_i \mid O_c). \tag{3.11}$$

This particular formulation and also the independence assumption of the occupancy grid makes this framework a perfect candidate for highly efficient parallel computing implementations of exact inference – for more information on this subject, please refer to the worked out examples of Part II and also Appendix A.

3.4.2 Visuoauditory Sensor Models for Occupancy Grids

One of the beauties of the Bayesian approach is the fact that very disparate sensory modalities can be modelled so that they can be used in a sensor fusion framework – with the advantage of explicitly and seamlessly dealing with the inherent uncertainty — by relating them via *a common variable encoding the final perceptual outcome.* This variable constitutes, therefore, a

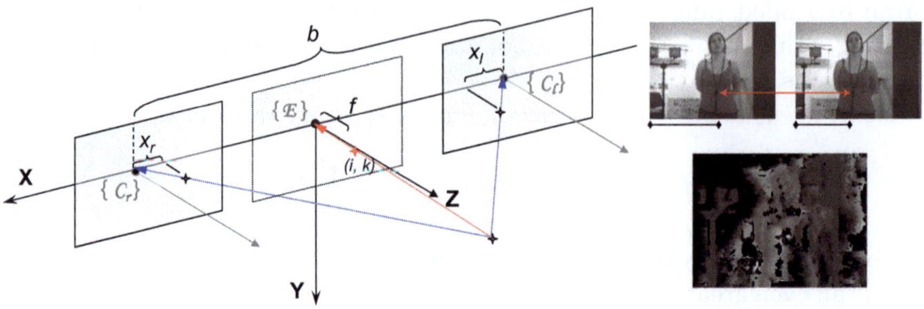

Fig. 3.7. Cyclopean geometry for stereovision. The Cyclopean view is the result of the combination of the images received from the two eyes or cameras in a stereovision setup, and is named after the mythical Cyclops who had only one eye. The use of Cyclopean geometry (pictured on the left for an assumed frontoparallel configuration) allows direct use of the egocentric reference frame for depth maps taken from the disparity maps yielded by the stereovision system (of which an example is shown on the right).

realisation of the *representation* of the surrounding environment investigated in Chapter 2. In the perspective of the roboticist trying to model multisensory perception, as a matter of fact, the objective is to find a useful perceptual representation for which models can be defined considering any available modality, irrespectively of the differing starting points represented by the actual sensor measurements.

In the following text, a published example of a sensor model for robotic vision will be defined using the concepts we have learned so far. This model can be used together with the robotic binaural sensor model presented in Example 2.3 so as to obtain a visuoauditory framework for perception. Note that sensor readings, expressed in both cases as Z_i, where i denotes the $i-th$ sensor of whatever kind, reflect completely different physical phenomena for sensing, and therefore as in most cases of sensor fusion have differing dimensions, units, etc. The common representation in these models is occupancy, denoted as O_c, and as such they are defined so as to relate Z_i in each of the corresponding modalities to the perception of the occupancy of a cell c.

Example 3.3. **A stereovision sensor model for occupancy grids**

As mentioned in Chapter 1, several authors argue that current evidence strongly suggests that the brain codes complex patterns of sensory uncertainty in its internal representations and computations. One such representation is believed to be neural population coding (e.g., average firing rate) — see Knill and Pouget [17]; Pouget et al. [25]; Jacobs [20]; Rao [16]; Zemel et al. [30]; Denève et al. [27]; Barber et al. [18], for just a few examples.

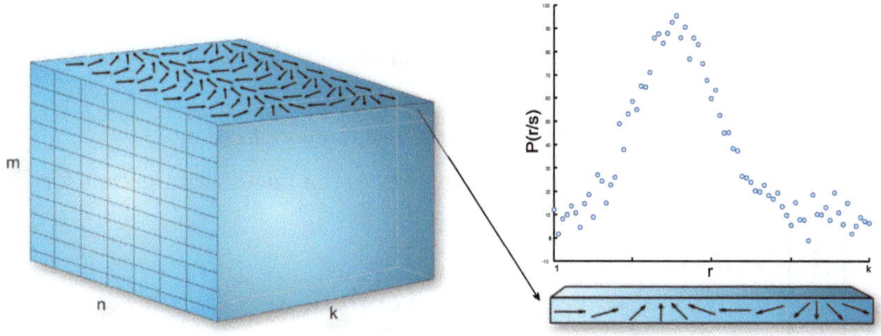

Fig. 3.8. Population code data structure. On the left, a spatially organised 2D grid has each cell (which might correspond, for example, to a specific area on the retina or a pixel on a digital image) associated to a population code simulation extending to a third dimension, represented on the right — i.e., a set of probability values of a neuronal population encoding a pdf (in this example, for preferred directions). Note that this map does not precisely mimic the cortical columnar architecture, and is just an approximation, and that the pdf can in fact extend to more than a single dimension (e.g., if the encoded property would be local velocity, two dimensions would be necessary so as to represent speed and direction).

Ferreira et al. [1, 8, 9] proposed a model based on a tentative data structure analogous to neuronal population activity patterns to represent uncertainty in the form of probability distributions [25]. Thus, a spatially organised 2D grid may have each cell (corresponding to a virtual photoreceptor in the Cyclopean view – see Fig. 3.7) associated to a "population code" extending to additional dimensions, yielding a set of probability values encoding a N-dimensional probability distribution (see Fig. 3.8). They decided to model these virtual photoreceptors in terms of their contribution to the estimation of cell occupancy in a similar fashion to the solution proposed by Yguel, Aycard, and Laugier [11]. This solution incorporates a complete formal definition of the physical phenomenon of occlusion (in this case, light reflecting from surfaces occluded by opaque objects do not reach the vision sensor's photoreceptors).

Let us start by defining the relevant indices, parameters and variables:

- Once a projection line (θ, ϕ), with $\theta_{\min} \leq \theta \leq \theta_{\max} \wedge \phi_{\min} \leq \phi \leq \phi_{\max}$, is established for a specific virtual photoreceptor, one may partition the respective line-of-sight into cells. Therefore, by abuse of notation and in order to simplify references to cells in the line-of-sight, these will be referred to using the abstraction $c \in \mathbb{N}, 1 \leq c \leq N$, where N denotes the total number of cells considered in the line-of-sight.
- O_c is a binary random variable denoting the occupancy state of cell c.

Program
- Description
  - Specification
    - Relevant variables:
      - $Z \in \{\text{"No Detection"}\} \cup \mathcal{Z}_{\text{VisDepth}}$: sensor depth measurement along line-of-sight (θ, ϕ);
      - O_c: binary value describing the occupancy of cell c;
      - $G_c \in \mathcal{G}_c \equiv \mathcal{O}^{N-1}$: state of all cells in the line-of-sight except for c.
    - Decomposition:
      $$P(Z \wedge O_c \wedge G_c) = P(O_c) \cdot \underbrace{P(G_c \mid O_c) P(Z \mid G_c \wedge O_c)}_{\text{Gives } P(Z \mid O_c) \text{ through } \sum_{G_c}.}$$
    - Parametric forms:
      - $P(O_c)$: uniform or prior estimate;
      - $P(G_c \mid O_c)$: unknown, apart from dependency on number of occupied cells;
      - $P(Z \mid G_c \wedge O_c)$: probability of a measurement by sensor, knowing first occupied cell is $k \equiv$ *elementary sensor model* $P_k(Z)$.
  - Identification:
    - Calibration for $P_k(Z) \Rightarrow P(Z \mid G_c \wedge O_c)$.
- Questions:
  - $P(Z \mid o_c)$
  - $P(O_c \mid z)$

Fig. 3.9. Bayesian program for vision sensor model of occupancy

- $G_c \in \mathcal{G}_c \equiv \mathcal{O}^{N-1}$ represents the state of all cells in the line-of-sight except for c. Each g_c is, thus, an $(N-1)$-tuple of the form $([O_1 = o_1], \cdots, [O_{c-1} = o_{c-1}], [O_{c+1} = o_{c+1}], \cdots, [O_N = o_N])$ given a specific cell c.
- $Z \in \{\text{"No Detection"}\} \cup \mathcal{Z}_{\text{VisDepth}}$ denotes a depth measurement along line-of-sight (θ, ϕ). More specifically, we are referring to a depth measurement that has been shifted, normalised and rounded so that the discrete values composing $\mathcal{Z}_{\text{VisDepth}}$ have a one-to-one correspondence to the cells on the line-of-sight. In other words, measurable space of Z excepting the case of "No Detection" is equivalent to the set of possible values for the cell index c.
- k denotes the *first occupied cell measured within the line of sight* (if there are other occupied cells from then onwards, they are occluded by k) and consequently the realisation of measurement Z by the virtual photoreceptor, $[Z = k]$. The way how the disparity map obtained by the stereovision rig and a potentially corresponding set of confidence values are actually related to k is explained further on.

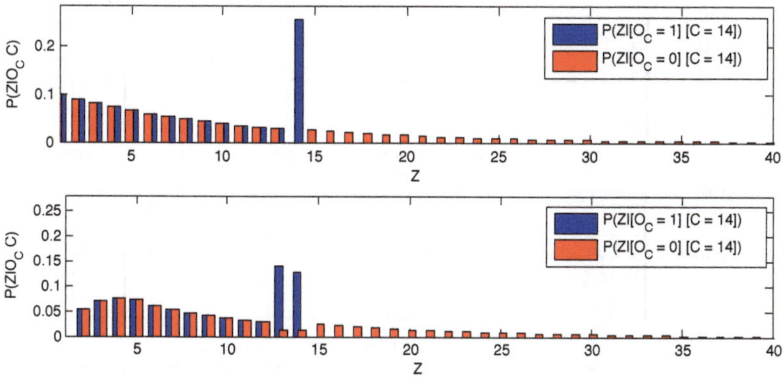

Fig. 3.10. Simulation results for direct vision sensor model $P(Z \mid O_c)$ for $c = 14$, given $P_{Empty} = .9$, $N = 40$, $\rho_{Min} = 1000\,\text{mm}$ and $\rho_{Max} = 11000\,\text{mm}$, considering both occupied and unoccupied states. Top: ideal sensor model (Dirac). Bottom: Gaussian elementary sensor model with $\sigma_\rho = 1\,\text{mm}$. Note that for the ideal sensor model, precision is maximal and aggregation is complete at $P([Z = 14] \mid [O_c = 1])$; additionally, note that for either of the presented cases, for $Z << 14$, $P(Z \mid [O_c = 1]) = P(Z \mid [O_c = 0])$, for $Z = 14$, $P(Z \mid [O_c = 1]) >> P(Z \mid [O_c = 0])$, and for $Z >> 14$, $P(Z \mid [O_c = 1]) \approx 0$, while $P(Z \mid [O_c = 0]) > 0$. This reflects the assumption coded in the model that, *when c is known to be occupied* (i.e. $[O_c = 1]$), cells farther from the origin than $c = 14$ are occluded, and hence do not yield visual readings.

The following expression gives the decomposition of the joint distribution of the relevant variables according to Bayes' rule and dependency assumptions:

$$P(Z \wedge O_C \wedge G_C) = P(O_c)P(G_c \mid O_c)P(Z \mid G_c \wedge O_c). \tag{3.12}$$

The parametric form and semantics of each component of the joint decomposition are then as follows:

- $P(O_c)$ represents *a priori* information on the environment. The probability of a cell being empty is $P_{\text{Empty}} = P([O_c = 0])$.
- $P(G_c \mid O_c) \equiv P(G_c)$ represents the probability that, knowing a state of a cell, the whole line-of-sight is in a particular state [11].
- $P(Z \mid G_c \wedge O_c)$ is sensor-dependent but, in any case, for all $(O_c, G_c) \in \mathcal{O} \times \mathcal{G}_c$, the probability distribution over Z *depends only on the first occupied cell, k*. Knowing the position of the first occupied cell in the projection line, $P(Z \mid G_k \wedge O_k)$ gives the probability of a measurement if k would be the only occupied cell in the line-of-sight. This particular distribution over Z is called the *elementary sensor model*, denoted by $P_k(Z)$.

The likelihood functions yielded by the population code data structure can then be formalised as

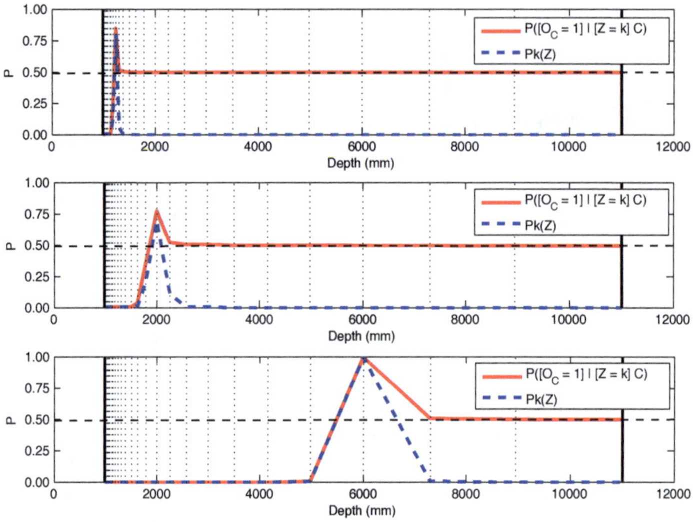

Fig. 3.11. Simulation results of inference to obtain $P(O_c \mid Z)$ using vision sensor model and log-partitioning of the line-of-sight. For all cases, $N = 40$, and $\rho_{Min} = 1000\,\text{mm}$ and $\rho_{Max} = 11000\,\text{mm}$ (delimited by full vertical lines), which results in $b \approx 1.2589\,\text{mm}$; each cell c is delimited by black, dashed vertical lines. In any of the graphs, the full red traces correspond to the result of inference (the horizontal axis in this case represents positions in depth throughout the line-of-sight and k the vision sensor measurement) and the full blue traces correspond to the Gaussian elementary sensor models (the horizontal axis in this case represents depth readings from the vision sensor and k the only occupied cell in the line-of-sight). Top: results for $\sigma_\rho = 20\,\text{mm}$, with $b^k + \rho_{Min} = 1200\,\text{mm}$. Middle and bottom: results for $\sigma_\rho = 100\,\text{mm}$, with $b^k + \rho_{Min} = 2000\,\text{mm}$ for the former and $b^k + \rho_{Min} = 5000\,\text{mm}$ for the latter. To note: the fact that Bayesian inference correctly yields the effects described originally by Elfes [34, 32], and the effects of the logarithmic partitioning of depth and of the soft evidence conveyed by the elementary sensor model.

$$P_k(Z) = L_k(Z, \mu_\rho(k), \sigma_\rho(k)), \begin{cases} \mu_\rho(k) & = \hat{\rho}(\hat{\delta}) \\ \sigma_\rho(k) & = \frac{1}{\lambda}\sigma_{min} \end{cases}, \tag{3.13}$$

a discrete probability distribution with mean μ_ρ and standard deviation σ_ρ, both a function of the cell index k, which directly relates to the distance ρ from the egocentric origin $\{\mathcal{E}\}$. Values $\hat{\delta}$ and λ represent the disparity reading and its correspondent confidence rating, respectively; σ_{min} and the expression for $\hat{\rho}(\hat{\delta})$ are taken from calibration, the former as the estimate of the smallest error in depth yielded by the stereovision system and the latter from the intrinsic camera geometry. The likelihood function *constitutes, in fact, the elementary sensor model* as defined above for each vision sensor, and formally represents *soft evidence* concerning the relation between vision sensor measurements denoted generically by Z and the corresponding readings $\hat{\delta}$ and λ, described by the calibrated expected value $\hat{\rho}(\hat{\delta})$ and standard deviation $\sigma_\rho(\lambda)$ for each sensor.

Equation (3.13) only partially defines the resulting probability distribution by specifying the random variable over which it is defined and an expected value plus a standard deviation/variance – a full definition requires the choice of a type of distribution that best fits the noisy probability values taken from the population code data structure. The traditional choice, mainly due to the central limit theorem, favours normal distributions $\mathcal{N}(Z, \mu_\rho(k), \sigma_\rho(k))$. Considering what happens in the mammalian brain, this choice appears to be naturally justified – biological population codes often yield bell-shaped distributions around a preferred reading [26; 15].

However, the fact that depth sensors always yield positive readings may be contradicted by the circumstance that normal distributions assign non-zero probabilities to negative depth values; even worse, close to the origin ($Z = 0$) this distribution assigns a *high* probability to negative depth values! With this purpose, Ferreira et al. adapted the Gaussian elementary sensor model by Yguel et al.

$$P_k([Z=z]) = \begin{cases} \int_{]-\infty;0]} \mathcal{N}(\mu(k-0.5), \sigma(\sigma_\rho))(u)du, & z \in [0;1] \\ \int_{\lceil z \rceil -1}^{\lceil z \rceil} \mathcal{N}(\mu(k-0.5), \sigma(\sigma_\rho))(u)du, & z \in]1;N] \\ \int_{]N;+\infty]} \mathcal{N}(\mu(k-0.5), \sigma(\sigma_\rho))(u)du, & z = \text{“No Detection”} \end{cases} \tag{3.14}$$

where $\mu(\bullet)$ and $\sigma(\bullet)$ are the operators that perform the required spatial coordinate transformations, and $k = \lceil \mu_\rho \rceil$ is assumed to be the index of the only occupied cell in the line-of-sight, which represents the coordinate interval $]k-1;k]$.

Camera calibration can be performed using a standard stereovision calibration software to estimate left and right camera *intrisic parameters* (i.e., focal length and distortion parameters for undistorting images for processing) and *extrinsic parameters* (i.e., transformation between camera local coordinate systems – in the case of an ideal frontoparallel setup, the estimation of baseline b) that allow the application of the reprojection equation

$$\begin{bmatrix} 1 & 0 & 0 & 0 \\ 0 & 1 & 0 & 0 \\ 0 & 0 & 0 & f \\ 0 & 0 & \frac{1}{b} & 0 \end{bmatrix} \begin{bmatrix} u_l - \frac{\hat{\delta}}{2} \\ v_l \\ \hat{\delta} \\ 1 \end{bmatrix} = \begin{bmatrix} WX \\ WY \\ WZ \\ W \end{bmatrix}, \tag{3.15}$$

where u_l is the horizontal coordinate and v_l is the vertical coordinate of a point on the left camera, and $\hat{\delta}$ is the disparity estimate for that point, all of which in pixels, f and b are the estimated focal length and baseline, respectively, both of which in metric distance, and X, Y and Z are 3D point coordinates respective to the egocentric/cyclopean referential system $\{\mathcal{E}\}$.

Using reprojection error measurements given by the calibration procedure, parameter σ_{min} of equation (3.14) is defined as being equal to the maximum error exhibited by the stereovision system.

Finally, to determine $(\theta_{i,k}, \phi_{i,k})$ and $\hat{\rho}_{i,k}(\hat{\delta})$ (i.e. to perform the Cartesian-to-spherical transformation) for each projection line (i,k) to use with the vision sensor model given in Figure 3.9, the following relations are built from equation (3.15),

$$\begin{cases} \theta_{i,k} & = 2\arctan(\frac{X}{2f}) \\ \phi_{i,k} & = 2\arctan(\frac{Y}{2f}) \\ \hat{\rho}_{i,k}(\hat{\delta}) & = \sqrt{X^2(\hat{\delta}) + Y^2(\hat{\delta}) + Z^2(\hat{\delta})} \end{cases} \tag{3.16}$$

Given $\theta_{i,k}$ and $\phi_{i,k}$, it becomes possible at any moment to compute depth from a given disparity estimate by substitution of the two first expressions onto the last in Equation 3.16, yielding

$$\hat{\rho}_{i,k}(\hat{\delta}) = f\sqrt{4\left(\tan^2\frac{\theta_{i,k}}{2} + \tan^2\frac{\phi_{i,k}}{2}\right) + \left(\frac{b}{\hat{\delta}}\right)^2} \tag{3.17}$$

The answer to the Bayesian program question in order to determine the sensor model $P(Z \mid O_c)$ for vision, which is in fact related to the decomposition of interest $P(O_c \wedge Z) = P(O_c)P(Z \mid O_c)$, is answered through Bayesian inference on the decomposition equation given in (3.12); the inference process will dilute the effect of the unknown probability distribution $P(G_c \mid O_c)$ through marginalisation over all possible states of G_c. In other words, the resulting *direct* model for vision sensors is based solely on knowing which is the first occupied cell on the line-of-sight and its relative position to a given cell of interest C (results of inference simulations are presented in Fig. 3.11).

To correctly formalise the Bayesian inference process, a formal auxiliary definition with respective properties follow.

Definition 1. $T_c^k \in \mathcal{G}_c$ *is the set of all tuples for which the first occupied cell is* k. *Formally, it denotes tuples such as* $(o_1, \cdots, o_{c-1}, o_{c+1}, \cdots, o_N) \in \{0,1\}^{N-1}$, *yielding* $[O_i = 0] \wedge [O_k = 1], \forall i < k$.

Property 1.1. $\forall (i,j), i \neq j, T_c^i \bigcap T_c^j = \emptyset$

Property 1.2. $\bigcup T_c^k = \mathcal{G}_c \setminus \mathcal{G}_\emptyset$, *with*
$\mathcal{G}_\emptyset = \{(o_p)_p \mid \forall p \in \mathbb{N} \setminus \{c\}, 1 \leq p \leq N, [O_p = 0]\}$

Property 1.3. *If* $k < c$ *there are* k *determined cells: the* $k-1$ *first cells,* $(o_1, \cdots, o_{k-1})$, *which are empty, and the* k*th,* (o_k), *which is occupied.* Then, $P(T_c^k) = P_{Empty}^{k-1}(1 - P_{Empty})$.

Property 1.4. *If* $k > c$ *there are* $k-1$ *determined cells: the* $k-2$ *first cells,* $(o_1, \cdots, o_{c-1}, o_{c+1}, \cdots, o_{k-1})$, *which are empty, and the* $(k-1)$*th,* (o_k), *which is occupied.* Then, $P(T_c^k) = P_{Empty}^{k-2}(1 - P_{Empty})$.

It now becomes possible to determine $P(Z \mid O_c)$ in order to express the desired joint distribution $P(Z \wedge O_c)$. This process leads to four distinct possible cases, that will be described next.

In the case of detection given an occupied cell c, the sensor measurement can only be due to the occupancy of this cell or a cell before it in terms of visibility. Thus [11],

$$\begin{aligned}
&\forall Z \neq \text{“No Detection”}, \\
&P(Z \mid [O_C = 1]) = \\
&\quad = \sum_{g_c \in \mathcal{G}_c} P([G_c = g_c]) P(Z \mid [O_c = 1] \wedge [G_c = g_c]) \\
&\quad = \sum_{k=1}^{c-1} P(T_c^k) P_k(Z) + (1 - \sum_{k=1}^{c-1} P(T_c^k)) P_c(Z) \\
&\quad = \sum_{k=1}^{c-1} P_{\text{Empty}}^{k-1} (1 - P_{\text{Empty}}) P_k(Z) + P_{\text{Empty}}^{c-1} P_c(Z)
\end{aligned} \tag{3.18}$$

Equation (3.18) has two terms: the left term that represents the case where c is occupied and the right term that comes from the aggregation of all the remaining probabilities around the last possible cell that might produce a detection: c itself. The “No Detection” case ensures that the distribution is normalised.

In the case of no detection given an occupied cell c, which would correspond most probably to the effects of occlusion from earlier cells,

$$\begin{aligned}
&Z = \text{“No Detection”}, \\
&P(Z \mid [O_c = 1]) = 1 - \sum_{r \neq \text{“No Det.”}} P([Z = r] \mid [O_c = 1])
\end{aligned} \tag{3.19}$$

In the case of a measurement from detection knowing that c is empty, where a erroneous detection is yielded by the sensor (the so-called *false alarm*),

$$\begin{aligned}
&\forall Z \neq \text{“No Detection”}, \\
&P(Z \mid [O_c = 0]) = \\
&\quad = \sum_{g_c \in \mathcal{G}_c} P([G_c = g_c]) P(Z \mid [O_c = 0] \wedge [G_c = g_c]) \\
&\quad = \sum_{k=1, k \neq c}^{N} P(T_c^k) P_k(Z) + P(\mathcal{G}_\emptyset) \delta_{Z = \text{“No Detection”}} \\
&\quad = \sum_{k=1}^{c-1} P_{\text{Empty}}^{k-1} (1 - P_{\text{Empty}}) P_k(Z) + \\
&\quad + \sum_{k=c+1}^{N} P_{\text{Empty}}^{k-2} (1 - P_{\text{Empty}}) P_k(Z) + P_{\text{Empty}}^{N-1} \delta_{Z = \text{“No Det.”}}
\end{aligned} \tag{3.20}$$

There are three terms in the empty cell, from left to right, corresponding respectively to before the detection, after the detection and no detection at all. Again, the "No Detection" case ensures that the distribution is normalised.

In the case of no detection knowing that c is empty, which will either be due to a miss-detection or a completely empty line-of-sight corresponding to $\mathcal{G}_{\emptyset}$,

$$\begin{aligned} &Z = \text{“No Detection”}, \\ &P(Z \mid [O_c = 0]) = \\ &\quad = 1 - (\sum_{r}^{N} P([Z = r] \mid [O_c = 0])) + P_{\text{Empty}}^{N-1} \delta_{Z=\text{“No Det.”}} \end{aligned} \tag{3.21}$$

The Bayesian program that summarises this model is presented on Fig. 3.9. Two questions are of particular interest to be answered by inference – the condensed version of the sensor model $P(Z \mid O_c)$, to be used in a hierarchical occupancy grid model, and the inverse sensor model $P(O_c \mid Z)$, that provides an estimate of the occupancy state of cells along the considered line-of-sight given a reading from a particular virtual photoreceptor (see Fig. 3.10 and Fig. 3.11 for respective simulation results).

The robotic binaural sensor model presented in Example 2.3 is defined independently of the actual cell geometry; therefore, one may choose any grid configuration in that case without loss of generality. On the other hand, both models are defined assuming egocentric referencing; consequently, registration and integration must be cared for if the final perception is to be related to allocentric frames of reference.

In the particular case of the stereovision model, the tesselation used is one-dimensional along the line-of-sight. If this type of model is to be used consecutively for all disparity values in a disparity map (in other words, the full set of virtual photoreceptors in the Cyclopean image are to be used), which is, of course, the natural course of action, then a higher dimensional grid will be used for the hierarchical occupancy grid model, most probably using a 3D tesselation. If a Cartesian grid is to be used, namely in conjunction with an allocentric reference frame, a ray-tracing method must be applied, followed by a registration algorithm so as to establish the correspondences of cells on the line-of-sight with the overall grid – refer to [11] to understand better what this entails in a 2D tesselation example.

On the other hand, if an egocentric, spherical representation is used, then *both* models can be used directly, without any modifications. Even if a log-partitioning of distance such as that of the BVM (Example 2.1) is used, it is easily shown that stereovision model dispenses any additional adaptation since it is generalisable in distance along the spherical direction of the respective line-of-sight. This type of application will be demonstrated in Part II of this book, Chapter 8.

The notions supporting the use both these models as plug-ins or *subroutines* in a hierarchical framework will be introduced in Chapter 4.

3.4.3 The Bayesian Occupancy Filter (BOF)

Coué, Pradalier, Laugier, Fraichard, and Bessière [14] and also Tay, Mekhnacha, Chen, Yguel, and Laugier [10] devised two versions of an important extension to the occupancy grid approach, by applying a Bayesian filter to capture the dynamics of object motion within the environment being represented.

Tay et al., in particular, have thereby relaxed the strong restriction that objects remain static through time by introducing an extra variable to the inference framework that encodes the probability of (local) motion of an object occupying a cell in instant t to a neighbouring cell in $t+1$, thus shifting occupancy from the former to the latter during that time-lapse. These authors managed to do this without compromising the feasibility of exact inference inherent to the original occupancy grid concept.

For these reasons, we will introduce this model as our following example. Once again, the model is defined independently of the actual cell geometry, and therefore the associated grid configuration becomes arbitrary.

Example 3.4. **A Bayesian filter to deal with the dynamics of occupancy grids – the Bayesian Occupancy Filter (BOF)**

Let us start by stating and defining the relevant variables:

- $C \in \mathcal{Y}$ is random variable denoting an index which simultaneously localises and identifies the reference cell in the grid $\mathcal{Y}$. It is also used as a subscript of most of the random variables defined in this text, so as to explicitly state their relation to cells in the grid.
- $A_C \in \mathcal{A}_C \subset \mathcal{Y}$ is a random variable that denotes the *hypothetical antecedent* cell of reference cell C. The set of allowed antecedents $\mathcal{A}_C$ of reference cell C is composed by the $N+1$ cells on the BVM grid from which an object might have moved from, within the time interval going from the previous inference step $t-1$ to the present time t. The number of possible antecedents of any cell is arbitrary, but should include cell C itself (which would represent the hypothesis of an object occupying the reference cell remaining still).
- O_C is a binary variable denoting the occupancy $[O_C = 1]$ or emptiness $[O_C = 0]$ of cell C; O_C^{-1} denotes the occupancy state of the effective antecedent of C, A_C, in the previous inference step, which will propagate to the reference cell as the object occupying a specific A_C is moved to C.

Program / Description / Specification

Relevant variables:

$C \in \mathcal{Y}$: indexes a cell on the grid;

A_C: identifier of the antecedents of cell C (stored as with C);

$Z_1, \cdots, Z_S \in \{\text{"No Detection"}\} \cup \mathcal{Z}$: *independent* measurements taken by S sensors;

O_C, O_C^{-1}: binary values describing the occupancy of cell C,
for current and preceding instants, respectively;

V_C: velocity of cell C,
discretised into $N+1$ possible cases $\in \mathcal{V} \equiv \{v_0, \cdots, v_N\}$.

Decomposition:

$$P(C \wedge A_C \wedge O_C \wedge O_C^{-1} \wedge V_C \wedge Z_1 \wedge \cdots \wedge Z_S) =$$

$$P(A_C)P(V_C \mid A_C)P(C \mid V_C \wedge A_C)P(O_C^{-1} \mid A_C)P(O_C \mid O_C^{-1}) \prod_{i=1}^{S} P(Z_i \mid V_C \wedge O_C \wedge C)$$

Parametric forms:

$P(A_C)$: uniform;

$P(V_C \mid A_C)$: histogram;

$P(C \mid V_C \wedge A_C)$: Dirac, 1 *iff* $c_{\log_b \rho} = a_{\log_b \rho} + v_{\log_b \rho}\delta t$, $c_\theta = a_\theta + v_\theta \delta t$ and
$c_\phi = a_\phi + v_\phi \delta t$ (*constant velocity assumption*);

$P(O_C^{-1} \mid A_C)$: probability of preceding state of occupancy given set of antecedents;

$P(O_C \mid O_C^{-1})$: defined through transition matrix $T = \begin{bmatrix} 1-\epsilon & \epsilon \\ \epsilon & 1-\epsilon \end{bmatrix}$,
where ϵ represents the probability of non-constant velocity;

$P(Z_i \mid V_C \wedge O_C \wedge C)$: *direct measurement model* for each sensor i, given by respective sub-BP.

Identification:

None.

Questions:

$$P(O_c \wedge V_c \mid z_1 \wedge \cdots \wedge z_S \wedge c) \rightarrow \begin{cases} P(O_c \mid z_1 \wedge \cdots \wedge z_S \wedge c) \\ P(V_c \mid z_1 \wedge \cdots \wedge z_S \wedge c) \end{cases}$$

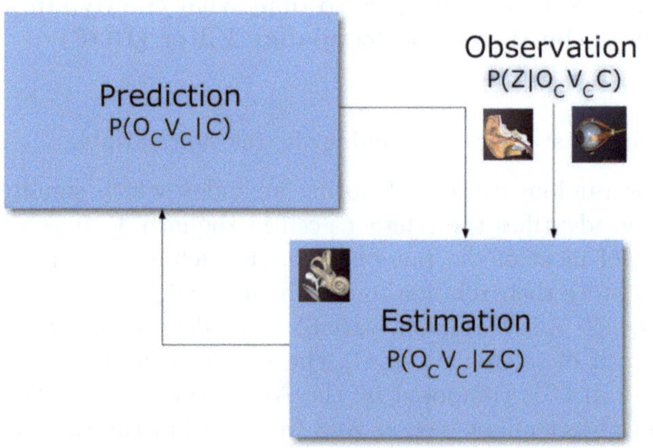

Fig. 3.12. Bayesian program for the estimation of Bayesian Volumetric Map current cell state (top), and corresponding Bayesian filter diagram (bottom – it considers only a single measurement Z for simpler reading, with no loss of generality). The respective filtering equation is given by (3.22) and (3.23), using two different formulations.

- V_C denotes the dynamics of the occupancy of cell C as a vector signalling local motion to this cell from its antecedents, discretised into $N+1$ possible cases for velocities $\in \mathcal{V} \equiv \{v_0, \cdots, v_N\}$, with v_0 signalling that the most probable antecedent of A_C is C, i.e. no motion between two consecutive time instants.
- $Z_1, \cdots, Z_S \in \{\text{"No Detection"}\} \cup \mathcal{Z}$ are *independent* measurements taken by S sensors.

The estimation of the joint state of occupancy and velocity of a cell is answered through Bayesian inference on the decomposition equation given in Fig. 3.12. This inference effectively leads to the Bayesian filtering formulation as used in the BOF grids.

Using the decomposition equation given in Fig. 3.12, given that $\prod_{i=1}^{S} P(Z_i \mid V_C \wedge O_C \wedge C)$ does not depend either on A_C or O_C^{-1}, we also have a more familiar formulation of the Bayesian filter,

$$\overbrace{P(V_C \wedge O_C \wedge Z_1 \wedge \cdots \wedge Z_S \wedge C)}^{\text{Estimation (Joint Distribution)}} = \overbrace{\prod_{i=1}^{S} P(Z_i \mid V_C \wedge O_C \wedge C)}^{\text{Observation}} \overbrace{\sum_{A_C, O_C^{-1}} P(A_C) P(V_C \mid A_C) P(C \mid V_C \wedge A_C) P(O_C^{-1} \mid A_C) P(O_C \mid O_C^{-1})}^{\text{Prediction}}. \tag{3.22}$$

Applying marginalisation and Bayes rule, we obtain the answer to the Bayesian program question, the global filtering equation

$$\overbrace{P(V_C \wedge O_C \mid Z_1 \wedge \cdots \wedge Z_S \wedge C)}^{\text{Estimation}} = \frac{\overbrace{\prod_{i=1}^{S} P(Z_i \mid V_C \wedge O_C \wedge C)}^{\text{Observation}} \overbrace{\sum_{A_C, O_C^{-1}} P(A_C) P(V_C \mid A_C) P(C \mid V_C \wedge A_C) P(O_C^{-1} \mid A_C) P(O_C \mid O_C^{-1})}^{\text{Prediction}}}{\underbrace{\sum_{A_C, O_C^{-1}, O_C, V_C} P(A_C) P(V_C \mid A_C) P(C \mid V_C \wedge A_C) P(O_C^{-1} \mid A_C) P(O_C \mid O_C^{-1}) \prod_{i=1}^{S} P(Z_i \mid V_C \wedge O_C \wedge C)}_{\text{Normalisation}}}. \tag{3.23}$$

Note that $P(Z_i \mid V_C \wedge O_C \wedge C)$ represents a direct sensor model for sensor i and might be, in fact, independent of either V_C or O_C (in which case it can be simplified to either $P(Z_i \mid O_C \wedge C)$ or $P(Z_i \mid V_C \wedge C)$, respectively), depending on what the roboticist has available. The visuoauditory sensor models defined in previously are examples of sensor models that might be used with this framework (provided that the appropriate registration and referencing issues are dealt with).

The process of solving the global filtering equation *by means of exact inference* can actually be separated into three stages, in practice. The first stage consists on the prediction of the probabilities of each occupancy and velocity state for cell $[C = c]$, $\forall k \in \mathbb{N}_0, 0 \leq k \leq N$,

$$\alpha_c([O_C = 1], [V_C = v_k]) = \sum_{A_C, O_C^{-1}} P(A_C) P(v_k \mid A_C) P(C \mid v_k \wedge A_C) P(O_C^{-1} \mid A_C) P(o_c \mid O_C^{-1}) \tag{3.24a}$$

$$\alpha_c([O_C = 0], [V_C = v_k]) = \sum_{A_C, O_C^{-1}} P(A_C) P(v_k \mid A_C) P(C \mid v_k \wedge A_C) P(O_C^{-1} \mid A_C) P(\neg o_c \mid O_C^{-1}), \tag{3.24b}$$

with o_c and $\neg o_c$ used as shorthand notations for $[O_C = 1]$ and $[O_C = 0]$, respectively.

The prediction step thus consists on performing the computations represented by (3.24) for each cell, essentially by taking into account the velocity probability $P([V_C = v_k] \mid A_C)$ and the occupation probability of the set of antecedent cells represented by $P(O_C^{-1} \mid A_C)$, therefore propagating occupancy states as a function of the velocities of each cell.

The second stage of the BVM Bayesian filter estimation process is multiplying the results given by the previous step with the observation from the sensor model, yielding, $\forall k \in \mathbb{N}_0, 0 \leq k \leq N$,

$$\beta_c([O_C = 1], [V_C = v_k]) = \prod_{i=1}^{S} (P(Z_i \mid v_k \wedge [O_C = 1] \wedge C)) \, \alpha_c([O_C = 1], v_k) \tag{3.25a}$$

$$\beta_c([O_C = 0], [V_C = v_k]) = \prod_{i=1}^{S} (P(Z_i \mid v_k \wedge [O_C = 0] \wedge C)) \, \alpha_c([O_C = 0], v_k), \tag{3.25b}$$

Performing these computations for each cell $[C = c]$ gives a non-normalised estimate for velocity and occupancy for each cell. The marginalisation over occupancy values gives the likelihood of each velocity, $\forall k \in \mathbb{N}_0, 0 \leq k \leq N$,

$$l_c(v_k) = \beta_c([O_C = 1], [V_C = v_k]) + \beta_c([O_C = 0], [V_C = v_k]). \tag{3.26}$$

The final normalised estimate for the joint state of occupancy and velocity for cell $[C = c]$ is given by

$$P(O_C \wedge [V_C = v_k] \mid Z_1 \wedge \cdots \wedge Z_S \wedge C) = \frac{\beta_c(O_C, [V_C = v_k])}{\sum_{V_C} l_c(V_C)}. \tag{3.27}$$

The related remaining questions of the BP for the BVM cell states, the estimation of the probability of occupancy and the estimation of the probability of a given velocity, are given through marginalisation of the free variable by

$$P(O_C \mid Z_1 \wedge \cdots \wedge Z_S \wedge C) = \sum_{V_C} P(V_C \wedge O_C \mid Z_1 \wedge \cdots \wedge Z_S \wedge C) \tag{3.28a}$$

$$P(V_C \mid Z_1 \wedge \cdots \wedge Z_S \wedge C) = \sum_{O_C} P(V_C \wedge O_C \mid Z_1 \wedge \cdots \wedge Z_S \wedge C). \tag{3.28b}$$

In summary, prediction propagates cell occupancy probabilities for each velocity and cell in the grid – $P(O_C \wedge V_C \mid C)$. During estimation, $P(O_C \wedge V_C \mid C)$ is updated by taking into account the observations yielded by the sensors $\prod_{i=1}^{S} P(Z_i \mid V_C \wedge O_C \wedge C)$ to obtain the final state estimate $P(O_C \wedge V_C \mid Z_1 \wedge \cdots \wedge Z_S \wedge C)$. The result from the Bayesian filter estimation will then be used for the prediction step in the next iteration.

Examples of the use of this model will be demonstrated in Part II of this book, Chapter 8.

3.5 Final Remarks and Further Reading

Hopefully, by now, the usefulness, power and elegance of probabilistic modelling for robotic perception will have become clear. This chapter served to introduce the basic tools for model construction and implementation; however, the reader is encouraged to delve deeper into whatever subject we have introduced in order to better undertake any project at hand.

For example, numerous references exist for anyone interested in Bayesian networks and any of its relatives or derivatives, namely the excellent introductory article by Charniak [33] and the general textbook by Jensen and Nielsen [12], while Bayesian programming was introduced by Bessière, Laugier, and Siegwart [7], and a comprehensive survey of models expressed using this formalism can be found in Diard, Bessiere, and Mazer [19]. Kalman filters were first introduced in the seminal publication by Kalman [36] and its variations discussed by Julier and Uhlmann [29]. Soft and hard evidence have been specifically covered by Valtorta, Kim, and Vomlel [21].

On the other hand, more complex inference techniques have only been summarily covered in this text, since they may be implemented in practice by resorting to APIs, software development kits and toolboxes such as the ones reviewed by Murphy [24] or Horn [2], and would therefore simply require the modeller to be aware of their existence. However, if the reader would be interested in, for example, working on a proprietary implementation based on any of these techniques, referral to the work by Ghahramani and Beal [22] or Minka [23] and other more advanced and/or specific texts is advised.

Finally, most of the issues introduced in the first two sections of this chapter and many other subjects are covered in detail, in the perspective of probabilistic graphical models, either by the seminal work by Pearl [35], or by textbooks such as the comprehensive manuscript by Koller and Friedman [5].

References

1. Ferreira, J.F., Castelo-Branco, M., Dias, J.: A hierarchical Bayesian framework for multimodal active perception. Adaptive Behavior 20(3), 172–190 (2012), doi:10.1177/1059712311434662 (Published online ahead of print, March 1) 87, 91
2. Horn, K.S.V.: (2012), `http://ksvanhorn.com/bayes/free-bayes-software.html` (retrieved in April 16, 2012) 83, 99
3. Colas, F., Diard, J., Bessiére, P.: Common Bayesian Models For Common Cognitive Issues. Acta Biotheoretica 58(2-3), 191–216 (2010) 77
4. Faria, D.R., Martins, R., Lobo, J., Dias, J.: Probabilistic Representation of 3D Object Shape by In-Hand Exploration. In: Proceedings of The 2010 IEEE/RSJ International Conference on Intelligent Robots and Systems, IROS 2010, Taipei, Taiwan (2010) XXII, 73, 74
5. Koller, D., Friedman, N.: Probabilistic graphical models: principles and techniques. MIT Press (2009) 100
6. Lunn, D., Spiegelhalter, D., Thomas, A., Best, N.: The BUGS project: Evolution, critique and future directions. Statistics in Medicine 28, 3049–3082 (2009) 83
7. Bessiére, P., Laugier, C., Siegwart, R. (eds.): Probabilistic Reasoning and Decision Making in Sensory-Motor Systems. STAR, vol. 46. Springer, Heidelberg (2008) ISBN 978-3-540-79006-8 72, 78, 79, 81, 99
8. Ferreira, J.F., Bessiére, P., Mekhnacha, K., Lobo, J., Dias, J., Laugier, C.: Bayesian Models for Multimodal Perception of 3D Structure and Motion. In: International Conference on Cognitive Systems (CogSys 2008), pp. 103–108. University of Karlsruhe, Karlsruhe (2008a) 87
9. Ferreira, J.F., Pinho, C., Dias, J.: Bayesian Sensor Model for Egocentric Stereovision. In: 14a Conferência Portuguesa de Reconhecimento de Padrões Coimbra, RECPAD 2008 (2008) 87
10. Tay, C., Mekhnacha, K., Chen, C., Yguel, M., Laugier, C.: An efficient formulation of the Bayesian occupation filter for target tracking in dynamic environments. International Journal of Autonomous Vehicles 6(1-2), 155–171 (2008) 95
11. Yguel, M., Aycard, O., Laugier, C.: Efficient GPU-based Construction of Occupancy Grids Using several Laser Range-finders. International Journal of Autonomous Vehicles 6(1-2), 48–83 (2008) 87, 89, 91, 93, 94
12. Jensen, F.V., Nielsen, T.D.: Bayesian networks and decision graphs. Springer (2007) 99
13. Mekhnacha, K., Ahuactzin, J.M., Bessiére, P., Mazer, E., Smail, L.: Exact and approximate inference in ProBT. Revue d'Intelligence Artificielle 21(3), 295–332 (2007) 83

14. Coué, C., Pradalier, C., Laugier, C., Fraichard, T., Bessiére, P.: Bayesian occupancy filtering for multitarget tracking: an automotive application. Int. Journal of Robotics Research 25(1), 19–30 (2006) 95
15. Born, R.T., Bradley, D.C.: Structure and Function of Visual Area MT. Annual Review of Neuroscience 28, 157–189 (2005), doi:10.1146/annurev.neuro.26.041002.131052 91
16. Rao, R.P.N.: Bayesian inference and attentional modulation in the visual cortex. NeuroReport — Cognitive Neuroscience and Neurophysiology 16(16), 1843–1848 (2005) ISSN 0899-7667 86
17. Knill, D.C., Pouget, A.: The Bayesian brain: the role of uncertainty in neural coding and computation. TRENDS in Neurosciences 27(12), 712–719 (2004) 86
18. Barber, M.J., Clark, J.W., Anderson, C.H.: Neural representation of probabilistic information. Neural Computation 15(8), 1843–1864 (2003), ISSN 0899-7667, doi:10.1162/08997660360675062 86
19. Diard, J., Bessiere, P., Mazer, E.: A survey of probabilistic models using the Bayesian programming methodology as a unifying framework. In: International Conference on Computational Intelligence, Robotics and Autonomous Systems (IEEE-CIRAS), Singapore (2003) 99
20. Jacobs, R.A.: What determines visual cue reliability? TRENDS in Cognitive Sciences 6(8), 345–350 (2002) Review 86
21. Valtorta, M., Kim, Y.G., Vomlel, J.: Soft evidential update for probabilistic multiagent systems. International Journal of Approximate Reasoning 29(71), 106 (2002) 99
22. Ghahramani, Z., Beal, M.J.: Propagation Algorithms for Variational Bayesian Learning. Neural Information Processing Systems 13 (2001) 82, 100
23. Minka, T.P.: Expectation Propagation for approximate Bayesian inference. In: UAI 2001, Proceedings of the 17th Conference in Uncertainty in Artificial Intelligence. Morgan Kaufmann Publishers Inc., San Francisco (2001) 82, 100
24. Murphy, K.: The Bayes Net Toolbox for Matlab. Computing Science and Statistics 33 (2001) 83, 99
25. Pouget, A., Dayan, P., Zemel, R.: Information processing with population codes. Nature Reviews Neuroscience 1, 125–132 (2000) Review 86, 87
26. Treue, S., Hol, K., Rauber, H.J.: Seeing multiple directions of motion — physiology and psychophysics. Nature Neuroscience 3(3), 270–276 (2000) 91
27. Denéve, S., Latham, P.E., Pouget, A.: Reading population codes: a neural implementation of ideal observers. Nature Neuroscience 2(8), 740–745 (1999), doi:10.1038/11205 86
28. Lebeltel, O.: Programmation Bayésienne des Robots. Ph.D. thesis, Institut National Polytechnique de Grenoble, Grenoble, France (1999) 78
29. Julier, S.J., Uhlmann, J.K.: A New Extension of the Kalman Filter to Nonlinear Systems. In: Kadar, I. (ed.) Signal Processing, Sensor Fusion, and Target Recognition VI. SPIE Proceedings, vol. 3068, pp. 182–193 (1997) 99
30. Zemel, R.S., Dayan, P., Pouget, A.: Probabilistic Interpretation of Population Codes. Advances in Neural Information Processing Systems 9, 676–683 (1997) 86
31. Buntine, W.L.: Operations for Learning with Graphical Models. Journal of Artificial Intelligence Research (AI Access Foundation) 2, 159–225 (1994) ISSN 11076-9757 73

32. Elfes, A.: Multi-Source Spatial Data Fusion Using Bayesian Reasoning. In: Abidi, M.A., Gonzalez, R.C. (eds.) Data Fusion in Robotics and Machine Intelligence. Academic Press (1992) 90
33. Charniak, E.: Bayesian networks without tears: making Bayesian networks more accessible to the probabilistically unsophisticated. AI Magazine 12(4), 50–63 (1991) 99
34. Elfes, A.: Using occupancy grids for mobile robot perception and navigation. IEEE Computer 22(6), 46–57 (1989) 90
35. Pearl, J.: Probabilistic Reasoning in Intelligent Systems: Networks of Plausible Inference, revised second printing edn. Morgan Kaufmann Publishers, Inc., Elsevier (1988) 82, 100
36. Kalman, R.E.: A New Approach to Linear Filtering and Prediction Problems. Transactions of the ASME - Journal of Basic Engineering 82, 35–45 (1960) 99

4

Hierarchical Combination of Bayesian Models and Representations

...any large computation should be split up into a collection of small, nearly independent, specialized subprocesses.

Vision, *David Marr (1982)*

The hierarchy of relations from the molecular structure of carbon to the equilibrium of the species and the ecological whole, will perhaps be the leading idea of the future.

Order and Life, *Joseph Needham (1923)*

4.1 Introduction

Ever since seminal work by Marr [11] and Fodor [10] up until more recent accounts such as given by Ballard [8] and many others on computational theories of perception and cognition, the link between the functional organization of perceptual sites in the brain and the underlying computational processes has led to the belief that *modularity* plays a major role in making these processes tractable. Modularity, in this sense, means that the flow of computation can be broken down into simpler processes. As a matter of fact, although the interconnections between these sites have increasingly been found to be much more intricate than Marr believed (including feedback and lateral links), the notion that the brain is organised in a modular fashion has been supported by countless findings in Neuroscience research, and is currently undisputed.

Hierarchical Bayesian methods are standard and powerful tools for analysing models and drawing inferences, and have been extensively applied in statistics, machine learning and throughout the empirical sciences [4; 1]. Hierarchical Bayesian methods provide the adequate framework for implementing modularity in perception. Firstly, these methods allow model development to take place at multiple levels of abstraction. Secondly, they offer the possibility of understanding emergent behaviour as resulting from a mixture of qualitatively and quantitatively different sources. And, thirdly, this framework is able to unify disparate models.

4.2 A Simple Hierarchical Bayesian Model

Theorists have had some difficulty in coming to terms with establishing the border between non-hierarchical and hierarchical models. To tackle this

J.F. Ferreira and J. Dias: *Probabilistic Approaches for Robotic Perception*, Springer Tracts in Advanced Robotics 91,
DOI: 10.1007/978-3-319-02006-8_4,

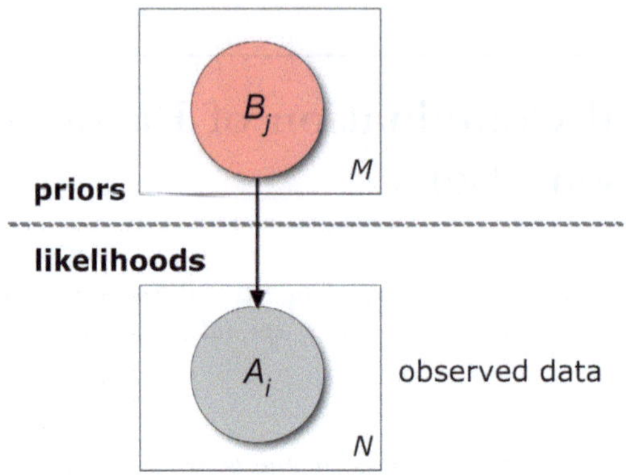

Fig. 4.1. Bayesian network for a non-hierarchical Bayesian model

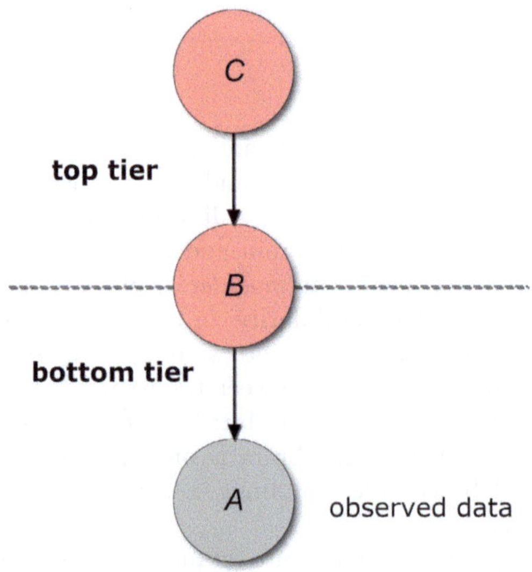

Fig. 4.2. Bayesian network for a simple two-tiered hierarchical Bayesian model. This hierarchy, which is also often called a *layered hierarchical model*, since each tier may be construed as a computational layer, can be generalised so as to involve an arbitrary number of tiers.

problem, we will attempt to simply and clearly offer our perspective on the definition of this border in formal terms.

Using an approach similar to Lee [1], we thereby define a *hierarchical Bayes model* as any generative model more complicated than the simplest type of model represented in Fig. 4.1. In the non-hierarchical model, sets of propositions represented by random variables A_i generate sets of observed data represented by random variables B_j; each of the generative variables A_i and observation variables B_j are conditionally independent within their respective set. This seemingly simple non-hierarchical type of model, which we say has only one tier represented graphically by the directed arc, encompasses many different useful existing types of model, some of which we addressed in previous chapters.

A graphical representation of a simple, generic, two-tiered hierarchical model is presented in Fig. 4.2. The corresponding decomposition is given by

$$\begin{aligned} P(A \wedge B \wedge C \mid \pi) &= P(A \mid \pi)P(B \mid A \wedge \pi)P(C \mid B \wedge A \wedge \pi) \\ &= P(A \mid \pi)P(B \mid A \wedge \pi)P(C \mid B \wedge \pi). \end{aligned} \tag{4.1}$$

Note that it is the fact that C is conditionally independent from A through B that confers the desired modularity to the model – we shall witness this in practice in the rest of this chapter.

Besides the question of modularity, the usefulness of hierarchical Bayesian models also relates to a well-understood technique in statistics – estimation via *shrinkage* [9]. The idea is that a naive estimate is improved by combining it with other information. Shrinkage is therefore used to further regularise ill-posed inference problems.

4.3 Building Hierarchies

4.3.1 Probabilistic Subroutines

Computer programming, especially in what concerns the so-called structured programming procedural paradigm, has always relied on the notion of modularity for constructing tractable computational solutions to complex problems, namely by introducing the subroutine concept. Adapting this concept to a Bayesian inference scenario, a *probabilistic subroutine* can be defined as a *submodel* that is exploited as a resource by a top-level model [2].

Consider a hierarchical Bayesian framework consisting of two models π_i, with $i = 1$ or 2, distinguished from one another through the use of context variable Π. Note that these two models have therefore different assumptions regarding the contexts given by their latent variables, as explained in Chapter 1, so this distinction makes sense. Moreover, the variables that they *do not explicitly share* will most certainly contribute to this difference, since they will be considered by the model in which they are not explicitly accounted for as hidden factors.

Let A and B be variables of interest for the top-level model, π_1, and imagine that the actual decomposition equation of this model is given by

$$P(A \wedge B \mid [\Pi = \pi_1]) = P(A \mid [\Pi = \pi_1])P(B \mid A \wedge [\Pi = \pi_1]).$$

Note that, when defining π_1, the model's context Π is known – putting it simply, the probability distributions of this model's decomposition equation are written given $[\Pi = \pi_1]$. Additionally, heed the fact that π_1, defined on its own, follows the non-hierarchical format portrayed on Fig. 4.1.

Now, what if the modeller has more sources of information that he or she would like to add to the model regarding the conditional dependence of the variables of interest, given in the previous expression by $P(B \mid A \wedge [\Pi = \pi_1])$, thereby increasing its power, but at the same time would like to minimise complexity?

The modeller could do this by resorting to modularity, for example by introducing a submodel, π_2, sharing A and B with top-level model π_1. If such a submodel would be at the modeller's disposal, then it would just be a matter of specifying the conditional dependence distribution as a call to a subroutine inferred from it, in the following fashion:

$$P(B \mid A \wedge [\Pi = \pi_1]) = P(B \mid A \wedge [\Pi = \pi_2]). \tag{4.2}$$

Note that, for the subroutine to be useful, the actual conditional dependence distribution should not be a factor in the decomposition of the joint probability distribution over all the variables of submodel π_2, or else it might as well have been specified directly when defining π_1. The distribution of interest $P(B \mid A \wedge [\Pi = \pi_2])$ should, on the contrary, be the result of inference using model π_2.

As a Bayesian Program, π_2 can be asked any probabilistic question related to its variables and can be arbitrarily complex. Moreover, many different models can question π_2, and π_2 itself might use other models as subroutines. Therefore, submodel π_2 obviously constitutes a resource to be exploited, and at the same time a potential exploiter of other resources [2]. On the other hand, the overall model is now hierarchical, as opposed to what would happen if π_1 would have been defined on its own.

Now, a more attentive reader might have noticed that there are models we have introduced in previous chapters that already use the subroutine concept, even beyond those we have signalled as doing so. An important, representative example would be dynamic Bayesian networks. In fact, when time-invariant, the model can be seen to be an application of the subroutine construct: in other words, its global model, given by (3.3), which deals with the complete time sequence, is a result of the iteration of the homogeneous local model [2].

4.3.2 *Probabilistic Conditional Weighting and Switching – Mixture Models*

What if Π, used previously to select between models by differentiation of contexts, becomes the focal point of the hierarchical framework? In other words, the actual context selection or even context blending process would now become an integral part of the inference mechanism through the decomposition equation

$$P(A \wedge B \wedge C \wedge \Pi \mid \pi_{\text{mixture}}) = \\ P(C \mid \pi_{\text{mixture}})P(A \mid \pi_{\text{mixture}})P(\Pi \mid C \wedge \pi_{\text{mixture}})P(B \mid A \wedge \Pi \wedge \pi_{\text{mixture}}),$$

where A and B are, again, the variables of interest.

This model constitutes a probabilistic conditional statement, where C represents the test condition, $P(C \mid \pi_{\text{mixture}})$ and $P(A \mid \pi_{\text{mixture}})$ are arbitrary priors (let us assume they are uniform distributions), the probability distribution $P(\Pi \mid C \wedge \pi_{\text{mixture}})$ represents a histogram of weights, and $P(B \mid A \wedge \Pi \wedge \pi_{\text{mixture}})$ is a collection of N models representing the conditional probability of B given A, $P(B \mid A \wedge [\Pi = \pi_i])$, with $i = 1 \ldots N$.

This generative model is then used to ask the question

$$P(B \mid A \wedge C \wedge \pi_{\text{mixture}}) = \sum_{\Pi} P(\Pi \mid C \wedge \pi_{\text{mixture}})P(B \mid A \wedge \Pi \wedge \pi_{\text{mixture}}), \tag{4.3}$$

yielding what is called a generalised *mixture model*, since it is a hierarchical Bayes model built from a weighted sum of component models.

So, how does one use such a hierarchical construct in practice? The best way is to consecutively implement the following steps:

1. Determine how many different weighting scenarios are to be considered and materialise these as the support of the test condition variable C.
2. Next, define probability distributions as histograms of weights for each model $[\Pi = \pi_i]$ given each scenario $[C = c]$ so as to specify $P(\Pi \mid C \wedge \pi_{\text{mixture}})$.
3. Finally, define each component model conditionally relating the two variables of interest, $P(B \mid A \wedge [\Pi = \pi_i] \wedge \pi_{\text{mixture}}) \equiv P(B \mid A \wedge [\Pi = \pi_i])$.

If the modeller decides that only one weighting scenario exists (i.e., the support of C is a singleton, which means that the corresponding proposition is always true), then C is no longer necessary[1] and the mixture model question (4.3) reduces to

$$\begin{aligned} P(B \mid A \wedge \pi_{\text{mixture}}) &= \sum_{i=1}^{N} P([\Pi = \pi_i] \mid \pi_{\text{mixture}})P(B|A \wedge [\Pi = \pi_i] \wedge \pi_{\text{mixture}}) \\ &= \sum_{i=1}^{N} w_i \times P_i(B \mid A), \end{aligned} \tag{4.4}$$

[1] And hence only steps 2 and 3 of the script for the practical construction of the mixture model are needed.

representing the principled Bayesian approach treatment of the more traditional, single-scenario version of the mixture model, with w_i representing the specific weight of each model indexed by i and specified by $P_i(B \mid A) \equiv P(B \mid A \wedge [\Pi = \pi_i])$.

A final remark: note that the use of a Dirac delta distribution for $P(\Pi \mid C \wedge \pi_{\text{mixture}})$ given a specific mixture scenario $[C = c]$ shifts the mixture model from a *model weighting* to a *model switching* paradigm [2].

4.3.3 Model Recognition

As was just seen, mixture models combine different component models into a single framework. Conversely, one might be interested in investigating which model of a collection of N competing models better explains observed data – a process called *model recognition* [2].

Let $\Delta = \{\delta_i\}$ be the variables corresponding to the observed data (with δ_i representing a variable for each datum[2]), and let Π be, as before, the context variable differentiating each model from the collection.

Model recognition is performed by performing inference on the decomposition given by

$$P(\Pi \wedge \Delta) = P(\Pi)P(\Delta \mid \Pi),$$

where $P(\Pi)$ is a prior on the various models of the collection and $P(\Delta \mid \Pi)$ is the probability of observing the data given the model (i.e., the *model likelihood*).

Assuming that the data is *independently and identically distributed* (the so-called *i.i.d. assumption*), $P(\Delta \mid \Pi) = \prod_{i=1}^{N} P(\delta_i \mid \Pi)$, where $P(\delta_i \mid \Pi)$ does not depend on the index i of each datum. For each model in the collection, π_j, the distribution $P(\delta_i \mid [\Pi = \pi_j])$ is a call to the probabilistic subroutine of that model, as described in section 4.3.1.

The question for model recognition is, consequently, given by

$$P(\Pi \mid \Delta) \propto P(\Pi) \prod_{i=1}^{N} P(\delta_i \mid \Pi). \tag{4.5}$$

Model recognition is hierarchical in the sense that it is in fact a model that reasons over a set of models [2].

4.3.4 Layered vs. Abstracted Hierarchies

Intuitively, one expects that hierarchical Bayes models will in one way or another fall into the category of constructs such as presented in Fig. 4.2 – the so-called *layered hierarchies* – with N tiers related by causality directed

---

2 Also called a *case* – see Chapter 6.

from top to bottom. However, hierarchical Bayes models may also be built by applying the notion of *abstraction*.

Abstraction can be achieved through one of two means: *variable abstraction* or *model abstraction*. The former is a way of indirectly creating a hierarchical dependence between two models via a variable in the top-level model that encodes the probability of another variable from the low-level model, taken from applying a decision process to the result of inference in the submodel. In other words, variable abstraction creates an implicit probabilistic subroutine through the use of soft evidence (Chapter 3).

Model abstraction, on the other hand, is much less of an expedient, generating a proper class of hierarchical models. Diard [3] presented an excellent tutorial on these models by providing a sort of definition-by-example that makes their power much easier to understand when comparing to layered hierarchies. In the following lines, we will offer a description that similar to this didactics.

Consider a robot trying to act on a world which it perceives using its sensors. Let S denote the conjunction of sensory variables relating to the perceptual system of a robot, and U be the conjunction of control variables relating to the robot's actuators. The simplest model describing action given perception would encode a completely reactive behaviour by promoting the direct inference of actuation given sensation $P(U \mid S \wedge \pi_{\text{reactive}})$ from the non-hierarchical construct

$$P(S \wedge U \mid \pi_{\text{reactive}}) = P(U \mid \pi_{\text{reactive}})P(S \mid U \wedge \pi_{\text{reactive}}).$$

Let us now add more cognitive power to our robot by introducing a variable T encoding a set of behaviours representing the robot's programming on what action to take in different contexts given the same set of sensory observations.

There are only a couple of useful ways of defining the decomposition of the joint distribution $P(S \wedge T \wedge U \mid \pi_{\text{A}})$. By now, we are sure that the reader is suspecting that one would be the layered hierarchical model π_A given by

$$P(S \wedge T \wedge U \mid \pi_A) = P(U \mid \pi_A)P(T \mid U \wedge \pi_A)P(S \mid T \wedge \pi_A). \tag{4.6}$$

This model reflects the most intuitive view of causality between the variables: actuation U reflects a behaviour T, which in turn is caused by sensation S.

The other possibility is the abstracted hierarchical model π_B given by

$$P(S \wedge T \wedge U \mid \pi_B) = P(T \mid \pi_B)P(U \mid T \wedge \pi_B)P(S \mid T \wedge \pi_B). \tag{4.7}$$

This model, clearly less intuitive in nature, involves the product $P(U \mid T \wedge \pi_B)P(S \mid T \wedge \pi_B)$, which, by applying Bayes' rule, can be rewritten as $P(U \wedge S \mid T \wedge \pi_B)$. With this notation, if T is the variable of interest it is not necessary to acknowledge the actual dependence between U and S,

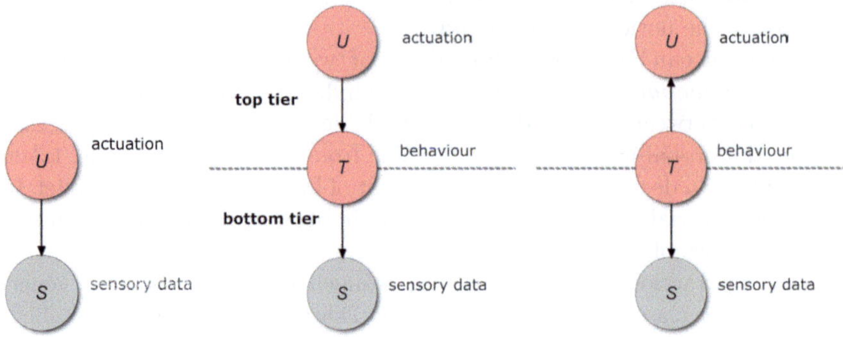

(a) Reactive model. (b) Layered model π_A. (c) Abstracted model π_B.

Fig. 4.3. Layered *vs* abstracted hierarchy for a robot trying to act on a world which it perceives using its sensors, comparing to a non-hierarchical purely reactive baseline model. The variables included in the Bayesian network representations given for each model are defined as follows: U represents actuation, S represents sensation, and T is an intermediate variable representing robot behaviour. Note that the layered nature of π_A is due to the fact that causation relationships flow consistently from top to bottom, as indicated by the directed arcs, as opposed to π_B, which is abstracted through the centralised dependence on T.

since this relationship is conditioned (i.e., *abstracted*) through T, even though this dependence is still encoded in the bottom-level model.

Let us now compare models π_A and π_B, shown in Fig. 4.3 side-by-side with the non-hierarchical framework, in terms of their strengths and weaknesses. The layered hierarchy allows for an easy interpretation of the inference of control given sensation, which is given for this model by

$$P(U \mid [S = s_i] \wedge \pi_A) \propto \sum_T P(T \mid U \wedge \pi_A) P([S = s_i] \mid T \wedge \pi_A), \qquad (4.8)$$

assuming a uniform distribution on actuation (i.e., no motor command is preferred over any other).

The interpretation of this inference process is as follows: for a given sensory input $[S = s_i]$, both the direct sensor model $P(S \mid T \wedge \pi_A)$ and the inverse actuator model $P(T \mid U \wedge \pi_A)$ are computed at the same time for all possible behaviours T, with the final result being the weighting of the several types of percept by the probability of the respective behaviour being enacted.

On the other hand, the abstracted hierarchy is much more appropriate for interpreting the inference of the actual behaviour being enacted, which is given by

$$\begin{aligned} P(T \mid [S = s_i] \wedge [U = u_i] \wedge \pi_B) \propto \\ P([S = s_i] \wedge [U = u_i] \mid T \wedge \pi_B) P(T \mid \pi_B). \end{aligned} \qquad (4.9)$$

The interpretation in this case is as follows: $P(S \wedge U \mid T \wedge \pi_B)$ models the influence of T on the sensorimotor relationship between U and S. In other words, in order to decide which behaviour T is to be enacted, the corresponding sensorimotor models are made to compete with one another, and the best behaviour will be the one which better predicts $[S = s_i] \wedge [U = u_i]$. Note that this is, in fact, model recognition as presented in section 4.3.3, assuming T as the model specifier.

Additionally, the power of the abstracted hierarchy becomes even more apparent when interpreting the inference of an additional variable X depending only on T, which is given by

$$P(X \mid [S = s_i] \wedge [U = u_i] \wedge \pi_B) \propto \sum_T P(X \mid T \wedge \pi_B) \underbrace{P([S = s_i] \wedge [U = u_i] \mid T \wedge \pi_B) P(T \mid \pi_B)}_{\text{model recognition}}. \quad (4.10)$$

In this case, the model recognition process weights the terms $P(X \mid T \wedge \pi_B)$, and therefore T serves the same role here as in the layered framework, except for the fact that now it adds more depth to the model by relating to *both* sensory and motor variables.

4.4 Examples of Hierarchical Bayes Model Applications

As a first example of an application of hierarchical modelling, we will present the basic measurement beam model introduced by Thrun, Burgard, and Fox [6], that the authors describe as "an approximate physical model of range finders". Consequently, this approach attempts to address, one by one, the most important phenomena with physical implications, and then generate a composite sensor model resulting from the mixture of the implications of each phenomenon.

Example 4.1. **A mixture model approach to an approximate physical beam model for range finders**

Range readings may be measured along a ray or beam – which would produce adequate assumptions for focussed sensors such as a laser range finder or photoreceptors, such as explained in Example 3.3 – or within a cone – a preferable approach when dealing with dispersive sensors, such as sonar [6]. As the title of the example suggests, we will be targeting the former, based on the description presented by Thrun et al. [6], although the model is easily generalisable.

The mixture hierarchy models the effects of four types of core measurement errors; we will address each one individually in the points that follow. The sensor model we will be modelling will be denoted as $P_k(Z \mid \pi)$, where Z denotes a range reading measured along the beam, π represents the context of

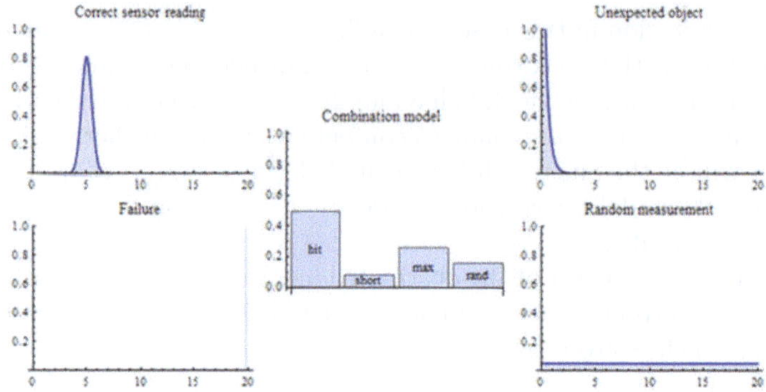

(a) Distributions corresponding to each of the components are represented around the mixture model weighting distribution.

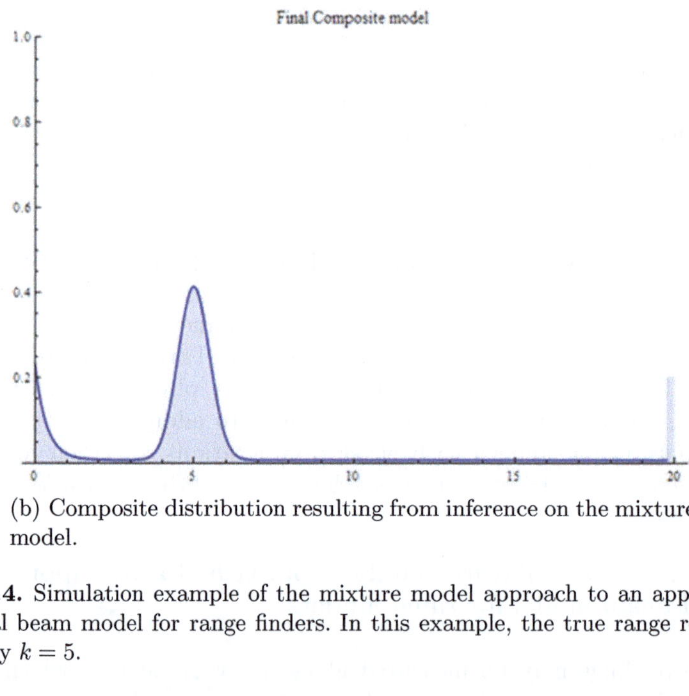

(b) Composite distribution resulting from inference on the mixture model.

Fig. 4.4. Simulation example of the mixture model approach to an approximate physical beam model for range finders. In this example, the true range reading is given by $k = 5$.

the beam (e.g., the beam's origin and orientation in space) and the component model being addressed, and k denotes the *true* range reading for the object being detected by the sensor – note that this corresponds to the *elementary sensor model* notion, as defined in Example 3.3.

1. **Correct range reading with local measurement noise.** Ideally, a range finder will always yield the correct range reading to the nearest object within its line-of-sight. However, even if the sensor's reaction is properly elicited by the object in its line-of-sight, the actual reading that

it will yield will be unavoidably be influenced by several unaccountable factors construed as noise (Chapter 1).
In this case, the elementary model may be expressed as presented in Example 3.3, Chapter 3. For the sake of simplicity, however, we will formulate this model here so as to assume a truncated normal in a continuous representation of the line-of-sight,

$$P_k(Z \mid \pi_{\text{hit}}) = \begin{cases} \eta_{\text{hit}} \mathcal{N}\left(\mu = k, \sigma_{\text{hit}}\right), & \text{if } 0 \leq Z \leq z_{\max}, \\ \\ 0, \quad \text{otherwise}, \end{cases}$$

where η_{hit} is a normalising constant, evaluating to

$$\eta_{\text{hit}} = \left(\int_0^{z_{\max}} \mathcal{N}\left(\mu = k, \sigma_{\text{hit}}\right) dZ \right)^{-1}.$$

2. **Unexpected objects.** Most probabilistic mapping makes the *static state* (i.e., all objects in the world are stationary) assumption. This assumption incorporates in the model the fact that it is unlikely – although not impossible – that the world state will change; therefore, very often any unexpected state change will be integrated into the overall uncertainty of the model, as being the effect of a phenomenon which is unaccounted for (as shown in Chapter 1, this is one of the key advantages of probabilistic approaches). But, of course, we can take this integration a step further, by explicitly, albeit simplistically, accounting for it by assuming that objects unexpectedly appearing within the line of sight might "corrupt" a range reading that was being taken, and hence yield range readings which are closer than k (but not farther, due to occlusion).
Mathematically, such situations may be described through the use of an *exponential distribution*, as follows

$$P_k(Z \mid \pi_{\text{short}}) = \begin{cases} \eta_{\text{short}} \lambda_{\text{short}} e^{-\lambda_{\text{short}} Z}, & \text{if } 0 \leq Z \leq k, \\ \\ 0, & \text{otherwise}, \end{cases}$$

with

$$\begin{aligned} \eta_{\text{short}} &= \left(\int_0^{k} \lambda_{\text{short}} e^{-\lambda_{\text{short}} Z} dZ \right)^{-1}, \\ &= \frac{1}{1 - e^{-\lambda_{\text{short}} k}}). \end{aligned}$$

3. **Failures.** Sometimes, objects in the line-of-sight are missed altogether. In the case of light-dependent sensors such as laser range finders, this might happen for all sorts of reasons: reflections, transparency, absorptions, etc.

The two typical outcomes of such failures are the {"No Detection"} case (i.e., no reading) and the maximum allowable reading $z_{\max}$ (which, in a proper situation, would arise only when no object is present within the full detectable range of the beam).
Considering, for simplicity sake, that the range sensor *always* returns a measurement value, and therefore the {"No Detection"} case does not exist, and knowing that, in practice, these situations are rather common, we will model this situation with a narrow, locally uniform distribution limited by z_max

$$P_k(Z \mid \pi_{\max}) = \begin{cases} \eta_{\max}, & \text{if } z_{\max} - \Delta \leq Z \leq z_{\max}, \\ 0, & \text{otherwise,} \end{cases}$$

with

$$\eta_{\max} = \frac{1}{\Delta}.$$

4. **Random measurements.** Finally, any sensor can occasionally produce inexplicable and apparently arbitrary measurements. Since there is virtually no way of determining the source and nature of such readings, it is common to model the corresponding uncertainty as a uniform distrbution spread over the entire measurement range,

$$P_k(Z \mid \pi_{\text{rand}}) = \begin{cases} \eta_{\text{rand}}, & \text{if } 0 \leq Z \leq z_{\max}, \\ 0, & \text{otherwise,} \end{cases}$$

with

$$\eta_{\text{rand}} = \frac{1}{z_{\max}}.$$

The final composite model is a mixture model, represented by the following particular case of the weighted sum of Equation (4.4)

$$\begin{aligned} P_k(Z \mid \pi_{\text{full}}) = & \underbrace{w_{\text{hit}} \times P_k(Z \mid \pi_{\text{hit}}) + w_{\text{short}} \times P_k(Z \mid \pi_{\text{short}})}_{\text{effectively depend on the value of } k} \\ & + \underbrace{w_{\max} \times P_k(Z \mid \pi_{\max}) + w_{\text{rand}} \times P_k(Z \mid \pi_{\text{rand}})}_{\text{always the same; do not depend on the value of } k}, \end{aligned}$$

with $w_{\text{hit}} + w_{\text{short}} + w_{\max} + w_{\text{rand}} = 1$.

The probability densities of the individual components taken for a specific value of k (note that not all the components actually depend on k) and an illustrative set of parameter values, together with the full beam mixture model with a particular set of weights $\{w_{\text{hit}}, w_{\text{short}}, w_{\text{max}}, w_{\text{rand}}\}$ are shown in Fig. 4.4.

This model is easily formulated using discrete values for Z (in which case, $P_k(Z \mid \pi_{\text{max}})$ would much more naturally be defined as a Dirac delta distribution centred on z_{max}). Also note that the first, third and fourth cases are integrated in the model presented in Example 3.3 (the former and the latter explicitly), and that the second case is not needed for the BOF framework of Example 3.4, given that the non-stationary assumption is already accounted for by the filter.

Finally, note that if this model is applied to a Cartesian grid representation, a ray-tracing algorithm must be applied to follow the beam, and most probably a different discretisation of the beam is needed for every given direction; however, if it is applied to the BVM configuration, presented in Example 2.1, no ray-tracing is needed and the number of partitions along the line-of-site is always the same.

The values for the parameters and weights of the beam model can either be preprogrammed by the robot designer, or learnt from calibration data, using the methods described in Chapter 6. Simulated results are presented in Fig. 4.4.

Next, we present an example consisting of an adaptation of an abstracted hierarchical model applied by Körding et al. [5] to the domain of human multimodal perception, under the name of *causal inference*.

Example 4.2. **An abstracted hierarchical model for robotic multisensory perception**

Perceptual cues are seldom contextually relevant by themselves, but rather acquire their significance through their meaning about their causes. Consequently, it is important to understand how cues from multiple sensory modalities can be used to infer the underlying causes better than the conclusions that can be derived from a single type of sensor.

In humans, the nervous system is constantly engaged in combining uncertain information from different sensory modalities into an integrated understanding of the causes of sensory stimulation. Over the last decade, many scientists have gone back to a probabilistic interpretation of cue combination, as we implicitly mention in Chapters 1 and 2.

In the case of visuoauditory perception, let us assume that there is a single variable in the outside world (e.g., the position of a person) that is causing the cues (auditory and visual information). Due to the uncertainty and possible ambiguity inherent to this scenario, each of the cues is assumed to be a noisy observation of this underlying variable. Due to noise in sensation, there is some uncertainty about the information conveyed by each cue and Bayesian

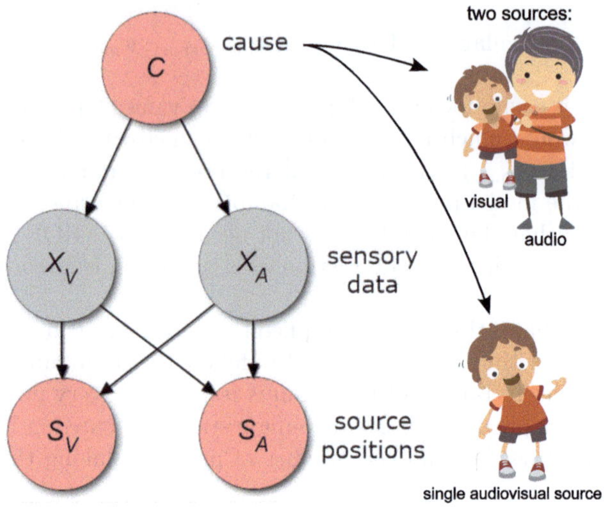

Fig. 4.5. Bayesian network for the abstracted hierarchical model for robotic azimuthal visuoauditory perception and respective submodels. Sensory variables are denoted as X, source position state variables are written as S and the abstracting variable signalling the actual cause for the visual and auditory percepts is represented by C, with $[C = 1]$ denoting a single source and $[C = 2]$ denoting two independent sources. These two possible causes are represented as a puppet and a puppet together with the puppeteer, respectively, illustrating the so-called *ventriloquist effect*, which describes the phenomenon of an observer perceiving the voice of the ventriloquist as being projected by the puppet. Subscripts V and A indicate visual and auditory modalities, respectively, in both sensory and source position state variables.

inference has been shown to be the systematic way of predicting how subjects could optimally infer position from the cues, if a single cause is assumed (see Chapter 2).

However, a range of experiments have shown effects that are hard to reconcile with the single-cause (i.e., forced-fusion) idea. In fact, auditory-visual integration breaks down when the spatial or temporal difference between the presentation of the visual and the auditory stimulus is large. Increasing this difference or inconsistency, for example by moving an auditory stimulus farther away from the position of a visual stimulus, reduces the influence each stimulus has on the perception of the other [5].

For this reason, Körding et al. [5] proposed an abstracted hierarchical model, shown on Fig 4.5, that weights the possibility of having either one or two sources eliciting the visual and auditory sensations and attempts to infer a percept from this process that would be coherent with human reports in a psychophysical study. Throughout this example, which will serve to apply Körding et al.'s model to a robotic perception scenario, the robot will only consider spatial distance along the azimuthal axis at a given instant, as in the original paradigm.

The objective of the model of Fig 4.5 is to infer an estimate for the azimuthal position state of the sources, denoted by S_V and S_A, given the observed independent sensor readings x_V and x_A, with subscripts V and A referring to the visual and to the auditory modalities, respectively.

The top tier of the model is, in fact, itself an abstracted hierarchical construct, which serves to infer what are the causes of sensory stimulation through

$$P(C \mid X_V \wedge X_A \wedge \pi) \propto P(X_V \wedge X_A \mid C \wedge \pi) P(C \mid \pi),$$

with C, the abstracting variable, denoting the causes for sensory stimulation: either $[C = 1]$ and a single source exists, and therefore $S_V = S_A = S$, or $[C = 2]$ and two independent sources exist, and consequently $S_V \neq S_A$.

As described in section 4.3.4, the two causal models corresponding to the different values of C will compete in terms of the appropriateness of each hypothesis on the causes in explaining the observations to decide between each corresponding version of the perceptual submodel $P(S_V \wedge S_A \mid C \wedge \pi)$.

The competing causal models $P(X_V \wedge X_A \mid C \wedge \pi)$ for each possible cause C, knowing that X_V and X_A are independent observations, are given by

$$\begin{aligned} &P(X_V \wedge X_A \mid [C = 1] \wedge \pi) = \\ &\sum_S P(X_V \mid S \wedge \pi) P(X_A \mid S \wedge \pi) P(S \mid \pi), \end{aligned}$$

since in this case $S_V = S_A = S$, and

$$\begin{aligned} &P(X_V \wedge X_A \mid [C = 2] \wedge \pi) = \\ &\sum_{S_V} P(X_V \mid S_V \wedge \pi) P(S_V \mid \pi) \times \sum_{S_A} P(X_A \mid S_A \wedge \pi) P(S_A \mid \pi), \end{aligned}$$

given that, here, S_V and S_A represent independent sources.

The result of inference on the abstracted model is subsequently used to weight the corresponding perceptual models in the final inference step, as follows

$$P(S_V \mid x_V \wedge x_A \wedge \pi) = \sum_C P(C \mid x_V \wedge x_A \wedge \pi) P(S_V \mid C \wedge x_V \wedge x_A \wedge \pi),$$

$$P(S_A \mid x_V \wedge x_A \wedge \pi) = \sum_C P(C \mid x_V \wedge x_A \wedge \pi) P(S_A \mid C \wedge x_V \wedge x_A \wedge \pi).$$

Körding et al. defined the likelihoods used in the decompositions as

$$P(X_V \mid S_V \wedge \pi) \equiv P(S_V \mid [C=2] \wedge x_V \wedge x_A \wedge \pi) \equiv \mathcal{N}(\mu_V = x_V, \sigma_V),$$
$$P(X_A \mid S_A \wedge \pi) \equiv P(S_A \mid [C=2] \wedge x_V \wedge x_A \wedge \pi) \equiv \mathcal{N}(\mu_A = x_A, \sigma_A),$$

and

$$\begin{aligned} P(X_V \mid S \wedge \pi) &\equiv P(S_V \mid [C=1] \wedge x_V \wedge x_A \wedge \pi) \\ &\equiv P(X_A \mid S \wedge \pi) \equiv P(S_A \mid [C=1] \wedge x_V \wedge x_A \wedge \pi) \\ &\equiv \mathcal{N}\left(\mu = \frac{x_V \sigma_V + x_A \sigma_A}{\sigma_V + \sigma_A}, \sigma = \sqrt{\frac{\sigma_V^2 \sigma_A^2}{\sigma_V^2 + \sigma_A^2}}\right), \end{aligned}$$

where the standard deviations σ_V and σ_A encode the uncertainty inherent to visual and auditory processing of azimuthal position, respectively. In Körding et al.'s experiments, human observers were found to have a relatively precise visual system ($\sigma_V = 2.14 \pm 0.22°$) when comparing with the auditory system ($\sigma_A = 9.2 \pm 1.1°$), which agrees with most of the scientific findings in this respect.

Additionally, the authors modelled the bias of the human perception system towards estimating stimuli as being placed centrally on the azimuthal axis (i.e. straight ahead respective to the observer) by encoding it on the priors on the sensory and source position state variables as follows

$$P(S \mid \pi) \equiv P(S_V \mid \pi) \equiv P(S_A \mid \pi) \equiv \mathcal{N}(\mu_P = 0, \sigma_P),$$

where the standard deviation σ_P encodes the uncertainty inherent to this bias. In Körding et al.'s experiments, human observers were found to have a relatively modest prior estimating stimuli to be more likely located centrally ($\sigma_P = 12.3 \pm 1.1°$). Nevertheless, the prior does make a difference, having been found by the authors to positively influence the quality of the inferred estimates and the degree to which the model agrees with experimental results.

As a final remark, the fact that **all** distributions in the decomposition equations are defined as normal distributions allows the models to have analytical solutions for inference – please refer to [5] for further details. However, the actual distribution types and values of their respective parameters in a robotic perception context are, of course, a matter completely left to the modeller to decide upon.

4.5 Final Remarks and Further Reading

The use of Bayesian hierarchies for robotic perception is still in its infancy. However, evidence such as presented by Lee and Mumford [7], which we strongly suggest reading, will undoubtedly fuel the development of more complex models using hierarchical constructs.

The reader is advised to read the discussions presented by Colas et al. [2] and also Diard [3], concerning hierarchical modelling itself, and by Lee [1] and Shiffrin et al. [4], specifically regarding hierarchical cognitive modelling.

References

1. Lee, M.D.: How cognitive modeling can benefit from hierarchical Bayesian models. Journal of Mathematical Psychology 55(1), 1–7 (2011); Special Issue on Hierarchical Bayesian Models 103, 104, 119
2. Colas, F., Diard, J., Bessiére, P.: Common Bayesian Models For Common Cognitive Issues. Acta Biotheoretica 58(2-3), 191–216 (2010) 105, 106, 108, 119
3. Diard, J.: Is your hierarchical Bayesian model layered, or abstracted? In: Proceedings for BACS Workshop on Hierarchies and loops (D 1.4), Bayesian Approach to Cognitive System, BACS (2009) 109, 119
4. Shiffrin, R.M., Lee, M.D., Kim, W., Wagenmakers, E.J.: A Survey of Model Evaluation Approaches With a Tutorial on Hierarchical Bayesian Methods. Cognitive Science 32, 1248–1284 (2008) 103, 119
5. Körding, K.P., Beierholm, U., Ma, W.J., Quartz, S., Tenenbaum, J.B., Shams, L.: Causal Inference in Multisensory Perception. PLoS ONE 2(9), e943 (2007), doi:10.1371/journal.pone.0000943 115, 116, 118
6. Thrun, S., Burgard, W., Fox, D.: Probabilistic robotics. MIT Press, Cambridge (2005) 111
7. Lee, T.S., Mumford, D.: Hierarchical Bayesian inference in the visual cortex. Journal of the Optical Society of America A 20(7), 1434–1448 (2003) 118
8. Ballard, D.H.: An Introduction to Natural Computation. MIT Press, Cambridge (1999) 103
9. McCallum, A., Rosenfeld, R., Mitchell, T., Ng, A.Y.: Improving text classification by shrinkage in a hierarchy of classes. In: Proceedings of the Fifteenth International Conference on Machine Learning, pp. 359–367 (1998) 105
10. Fodor, J.A.: The Modularity of Mind. MIT Press (1983) 103
11. Marr, D.: Vision: A Computational Investigation into the Human Representation and Processing of Visual Information. W. H. Freeman and Company, S. Francisco (1982) ISBN-13: 978-0716715672 103

5

Bayesian Decision Theory and the Action-Perception Loop

Perception is naturally surpassed toward action;
better yet, it can be revealed only in and through projects of action.
Being and Nothingness: An Essay on Phenomenological Ontology,
Jean-Paul Sartre (1943)

A real decision is measured by the fact that you've taken a new action.
If there's no action, you haven't truly decided.
Awaken the Giant Within, *Anthony Robbins (1992)*

Action is the real measure of intelligence.
unsourced quote, credited to Napoleon Hill (1883–1970)

5.1 Introduction

When presenting his enthralling talk on TED, Daniel Wolpert [2] put forth the following hypothesis on the evolutionary justification for the existence of the brain in Nature:

> I'm a neuroscientist. And in neuroscience, we have to deal with many difficult questions about the brain. But I want to start with the easiest question and the question you really should have all asked yourselves at some point in your life, because it's a fundamental question if we want to understand brain function. And that is, why do we and other animals have brains? Not all species on our planet have brains, so if we want to know what the brain is for, let's think about why we evolved one.
>
> Now you may reason that we have one to perceive the world or to think, and that's completely wrong. If you think about this question for any length of time, it's blindingly obvious why we have a brain. We have a brain for one reason and one reason only, and that's to produce adaptable and complex movements.
>
> There is no other reason to have a brain. Think about it. Movement is the only way you have of affecting the world around you. [...] So think about communication – speech, gestures, writing, sign language – they're all mediated through contractions of your muscles. So it's really important to remember that sensory, memory and

J.F. Ferreira and J. Dias: *Probabilistic Approaches for Robotic Perception*, Springer Tracts in Advanced Robotics 91,
DOI: 10.1007/978-3-319-02006-8_5,

cognitive processes are all important, but they're only important to either drive or suppress future movements.

Consequently, according to this perspective, any cognitive process, most of all perception, exists *only* to serve the promotion of physical action (i.e. *actuation*). On the other hand, from the point of view of robotics, this perspective is also particularly relevant, both due to the original and still core motivation of this field and due to the applications made possible by the current state-of-the-art robotics technology, if one acknowledges work as the most visible form of action (we refer the reader to the first paragraph of Chapter 1 regarding both accounts).

So, what do the concepts "action" and "actuation" really stand for in this context? The notion of action being defined as *operating a change of state*[1] might strike anyone as a rather trivial conclusion, but it is in fact pivotal in understanding perception and its purpose. As Wolpert so effectively points out, this change may come either from a direct physical modification of the world's state or by an indirect process, such as communication; it will, in any case, entail *actuation* and *motion* (i.e., "contraction of muscles", in the case of humans, mechanical actuation in the case of robots).

So, we come to the conclusion that perception ultimately exists in order to endow cognitive systems with the ability to derive a *decision* over which action, amongst a repertoire of actions available in a particular context, should be taken considering a given sensation. We go so far as to conjecture that *the decision process is, in fact, what makes humans intuit that their performance is deterministic, because for each instant in time a single decision arises concerning a particular situation, resulting in a single action.* In robotics, the decision process is often referred to as *planning*, while action is generally referred to as *control.*

This seems very much at odds with everything we have been building up upon since Chapter 3... The objective of perception thus far seemed to be to lead up to inference on a generative model, with which we wished to derive the posterior probability distribution resulting from the application of Bayes' rule, through a process that we call *probability propagation*. In effect, the result would be a collection of plausibility values concerning propositions, and definitively *not* a single value.

Therefore, in this chapter, we unveil the missing link in our story thus far. *Bayesian decision theory (BDT)*, which by many authors (see, for example, [18]) is described as the combination of probability theory with utility theory, provides a formal and complete framework for decision-making dealing with uncertainty. The probabilistic perspective to the action-perception loop resulting from applying Bayesian decision theory is illustrated in Fig. 5.1.

[1] See the formal definition for state, introduced in Chapter 3.

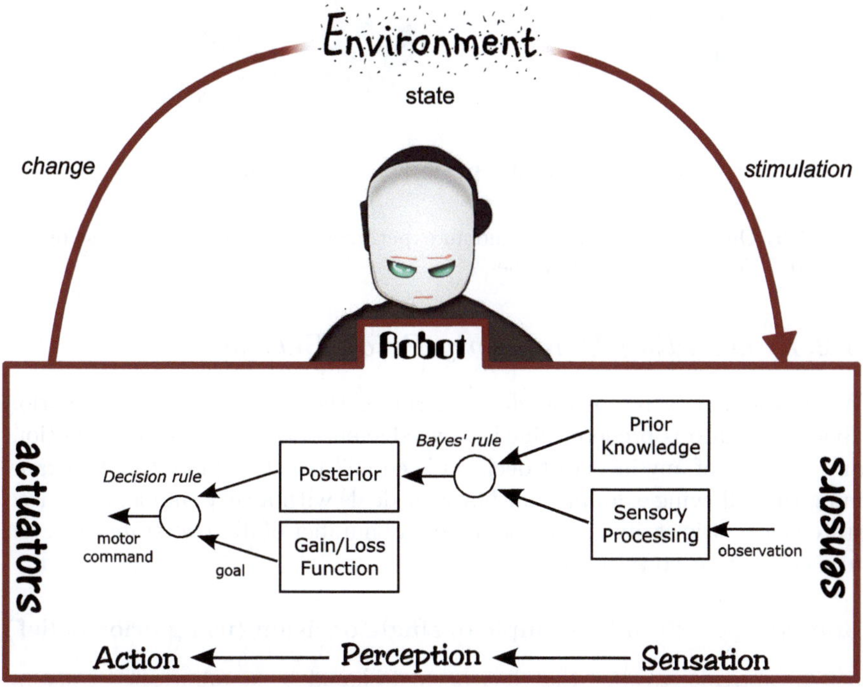

Fig. 5.1. Bayesian decision theory perspective on the action-perception loop (adapted from [8])

5.2 Unfolding Single Decision Rules Dealing with Uncertainty

As we were discussing above, decision from the probabilistic perspective can be described as the process of choosing a value (i.e., a scalar or a vector) from a given probability distribution. The *decision rule* therefore prescribes what value is to be chosen from the distribution at a given instant in time.

Let us start by considering situations for which no special care is taken to account for previous decisions – only the robot's current assessment of the state of the world matters, even if this process is repeated through time. The decision rule might be just obtaining the value by randomly sampling the distribution followed by the variable – this is sometimes called a *random draw*. However, this would clearly be a suboptimal decision when considering uncertainty.

In this section, we will use a worked-out case-study decomposed into several implementation examples to demonstrate how *single decision rules might be optimally devised based on uncertainty.*

Fig. 5.2. Outcome space of the random experiment of a robot manually sampling balls and cubes from an object pile

5.2.1 Deciding Using only Prior Beliefs

What would be a reasonable decision rule if the only available information about the state is predetermined knowledge encoded in a prior distribution, and the cost of any incorrect decision is equal? Let us explore this by examining the following example – although it deals with a very simple, two-cased state space, it is easily generalisable to any number of discrete states, or even to a continuous state space.

Example 5.1. **Simple example of single decision (using prior belief)**

Let us imagine a robot that has been designed by a roboticist to use an artificial hand to grab an object from a pile which is previously known to contain *only* either balls or cubes. This restriction will be obviously used by the roboticist as prior knowledge on the environment assimilated by the robot's model of the world.

Additionally, the roboticist might consider that the robot is conducting a random experiment with outcome space $W \equiv \{w_1, w_2\}$, with w_1 and w_2 representing the outcome of extracting a ball or a cube, respectively — see Fig. 5.2. Hence, the problem that the roboticist faces in designing this robot's perceptual system is "If the robot takes an object randomly from the pile, how does it decide how to classify the object it is holding?"

Therefore, the roboticist might have programmed the robot to have prior knowledge on how probable it would be to take an object of either class off the pile. This would consist of an educated guess which could have been learned from the statistical description of the pile, i.e. the knowledge of how many balls and cubes exist within it. The robot would then implement:

$$\text{Decide} \begin{cases} w_1, & P(w_1) > P(w_2), \\ w_2, & \text{otherwise.} \end{cases}$$

The probability of error in making this decision would consequently be assumed by the robot to be given by

$$P(error) = \min\left[P(w_1), P(w_2)\right]$$

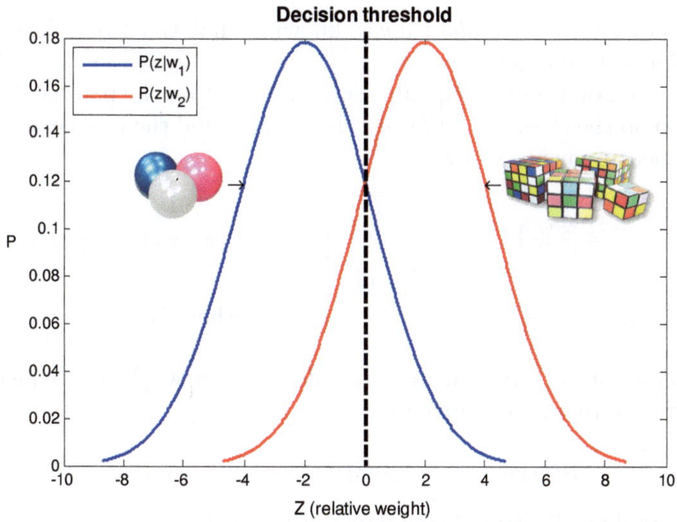

Fig. 5.3. Decision of a robotic perceptual system between a ball or a cube sampled from a pile using the likelihood function given by a manipulation sensor model. In this example, the sensor model is defined as a family of two normal distributions, one per class, and the standard reference mass was carefully chosen in such a way that the corresponding decision threshold falls at $Z = 0$. This means the robot will decide that the extracted object is a cube if $Z \geq 0$ and a ball otherwise.

This decision rule seems reasonable under such restrictive conditions... The downside is that it will always apply the same decision over and over again if the process is repeated, and if no further information is integrated into the model. On the other hand, if the prior is uniform, this rule will be ineffective. However, as we will see later on, under the given assumptions no other rule will perform any better!

5.2.2 Deciding Using the Likelihood Function – Maximum Likelihood Estimation (MLE)

So, what if the robot has access to sensors that allow it to observe the world's state, but no previous informative knowledge on it at the time of decision (i.e., a uniform prior)? Let us expand on our example, and continue our exploration journey.

Example 5.2. **Simple example of single decision (using likelihood)**

Let us now assume that the robot has weight measuring sensors on its artificial hand. The roboticist now develops a sensor model — a conditional distribution $P(Z \mid W)$ that gives the probability of measured weight values

$[Z = z]$ *relatively to a standard mass*, knowing that the object held by the robot is of a certain class.

If either class of objects is equally represented within the pile (i.e. the prior distribution on the classes is *uniform*), the robot would then implement, given a specific measurement $[Z = z]$

$$\begin{cases} P(w_1 \mid Z) = \frac{P(Z|w_1)}{P(Z)} \\ \\ P(w_2 \mid Z) = \frac{P(Z|w_2)}{P(Z)} \end{cases} \Rightarrow \text{Decide} \begin{cases} w_1, & P(z \mid w_1) > P(z \mid w_2), \\ \\ w_2, & \text{otherwise.} \end{cases}$$

The probability of error in making this decision would consequently be assumed by the robot to be given by

$$P(error \mid z) = \min\left[P(z \mid w_1), P(z \mid w_2)\right]$$

This example is illustrated concretely in Fig. 5.3.

Such a decision is said to be based on a *Maximum Likelihood Estimation (MLE) decision rule.*

5.2.3 Deciding Directly from Inference – Maximum a Posteriori Decision Rule (MAP)

We will now cover the case where both prior knowledge and sensory processing are available to the robot.

Example 5.3. **Simple example of single decision (from inference)**

The roboticist eventually decides that using prior knowledge on the pile *together* with a manipulation sensor model would allow building a much better model; certainly much more robust.

The robot would then implement, given a specific measurement $[Z = z]$

$$\begin{cases} P(w_1 \mid Z) = \frac{P(w_1)P(Z|w_1)}{P(Z)} \\ \\ P(w_2 \mid Z) = \frac{P(w_2)P(Z|w_2)}{P(Z)} \end{cases} \Rightarrow \text{Decide} \begin{cases} w_1, & P(w_1 \mid z) > P(w_2 \mid z), \\ \\ w_2, & \text{otherwise.} \end{cases}$$

The probability of error in making this decision would consequently be assumed by the robot to be given by

$$P(error \mid z) = \min\left[P(w_1 \mid z), P(w_2 \mid z)\right]$$

This example is illustrated concretely in Fig. 5.4.

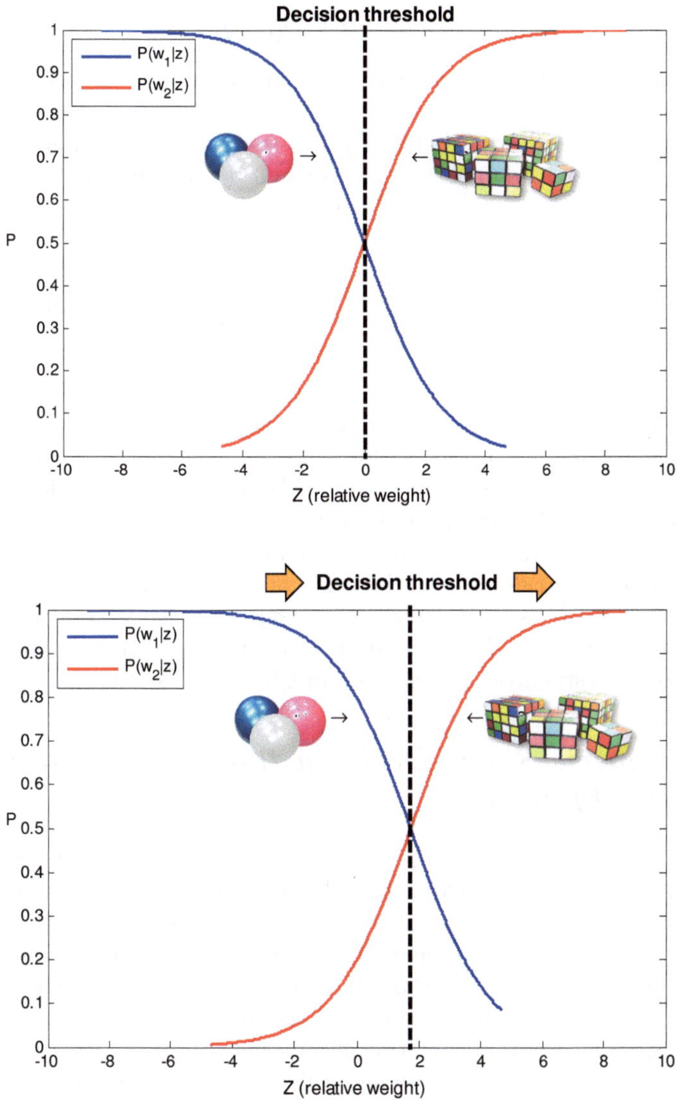

Fig. 5.4. Decision of a robotic perceptual system between a ball or a cube sampled from a pile using inference directly. Top: example using the same sensor model and the same standard reference mass as in Fig. 5.3, and a uniform prior distribution on the objects of the pile — this means the robot will decide that the extracted object is a cube if $Z \geq 0$ and a ball otherwise. Bottom: example using the same sensor model and the same standard reference mass as before, but now with a prior distribution on the objects of the pile given by $P(w_1) = .8$ and $P(w_2) = .2$ — this means that the decision threshold is moved to the right.

Let us generalise this situation through the following generative model

$$P(S \wedge O) = P(S)P(O \mid S), \tag{5.1}$$

where O and S are the observation and state variables, respectively, as in Chapter 3.

The *Maximum A Posteriori (MAP) decision rule* is used upon the result of inference on this model, performed as usual by applying Bayes' rule, and is given by

$$\begin{aligned} s^{\mathrm{MAP}} = \arg\max_S \left[P(S \mid O)\right] &= \arg\max_S \left[\frac{P(S)P(O \mid S)}{P(O)}\right] \\ &= \arg\max_S \left[P(S)P(O \mid S)\right]. \end{aligned} \tag{5.2}$$

Note that the evidence term $P(O)$ is not needed for the decision-making process, since it is only a normalising constant which does not depend on S. Furthermore, if all states have equal likelihood, then the decision will rely exclusively on the prior; conversely, if we have a uniform prior, then the decision will rely exclusively on the likelihood. This means that prior- or MLE-based decision rules are special cases of MAP. Finally, note that when the data used for decision is representative enough, the value derived from the use of either the MLE- or the MAP-based rules becomes a conventional statistical point estimator, as described in Chapter 1, section 1.2.4.

5.2.4 Generic Single Decision Rules – Assigning Utility/Risk

What if some perceptual outcomes, or actions resulting thereof, are not as desirable as others? In that case, the robot designer must define a *loss function*[2]. So let us say we have a set of N discrete perceptual states[3], so that $S \in \{s_1, s_2, \ldots, s_N\}$, but now we add to these a set of M discrete actions, so that $A \in \{a_1, a_2, \ldots, a_M\}$. The loss function $\Lambda(a_i \mid s_j)$ allows the robot to derive the loss it incurs when performing action a_i when the state is s_j.

Finally, we come to the formulation of the *generic decision rule using BDT*, as depicted in Fig. 5.1. When applying utility theory to deterministic models, the action to be chosen to implement would be the one that would minimise loss given a particular state (which the decision-maker would assume as being completely certain of about after observing it). The probabilistic approach to applying utility would, conversely, take full advantage of having probability values assigned to each state, and as such attempt to minimise the *expected loss given the observation*, generally referred to as *conditional risk*, expressed

[2] Or a reward or gain function, whichever makes more sense for the problem at hand, with no loss of generality.

[3] Again, easily generalisable to a continuous space.

Table 5.1. Zero-one loss function example for the ball vs cube decision. Using this loss function, $d_i = w_j$ (supposedly a correct decision) means that loss is zero, $d_i \neq w_j$ (supposedly an incorrect decision) that loss is one.

$\Lambda(d_i, w_j)$	d_1	d_2
w_1	0	1
w_2	1	0

as

$$R(a_i \mid O) = E\left[\Lambda(a_i \mid s_j)\right] = \sum_{j=1}^{M} \Lambda(a_i \mid s_j) P(s_j \mid O). \tag{5.3}$$

It is very important to understand that any generic Bayesian decision rule *will be optimal for its respective loss function.* Another important notion to grasp is that decision rules divide the observation space into decision regions separated by boundaries (in a one-dimensional context, the latter reduce to decision thresholds).

Let us now apply these concepts in order to bring closure to our worked-out example.

Example 5.4. **Simple example of single decision (by assigning utility/risk)**

Finally, the roboticist proceeds to program the robot by assigning a degree of risk incurred in deciding for either class of objects.

This is performed by assigning a risk value to each perceptual decision $D \equiv \{d_1 = \text{object is a ball}, d_2 = \text{object is a cube}\}$, and this in turn is achieved by defining a loss function $\Lambda(d_i \mid w_j)$.

Conditional risk (i.e., expected loss) would then provide the decision rule for the robot,

$$R(d_i \mid z) \Rightarrow \text{Decide} \begin{cases} d_1, & R(d_1 \mid z) < R(d_2 \mid z), \\ \\ d_2, & \text{otherwise.} \end{cases}$$

This example is illustrated concretely in Fig. 5.5.

As hinted earlier, conventional statistical estimators are obtained if actions are taken as an educated guess for the real value of the state (i.e., $a_j \equiv \hat{s}_i \approx s_i$), depending on the choice of the loss function.

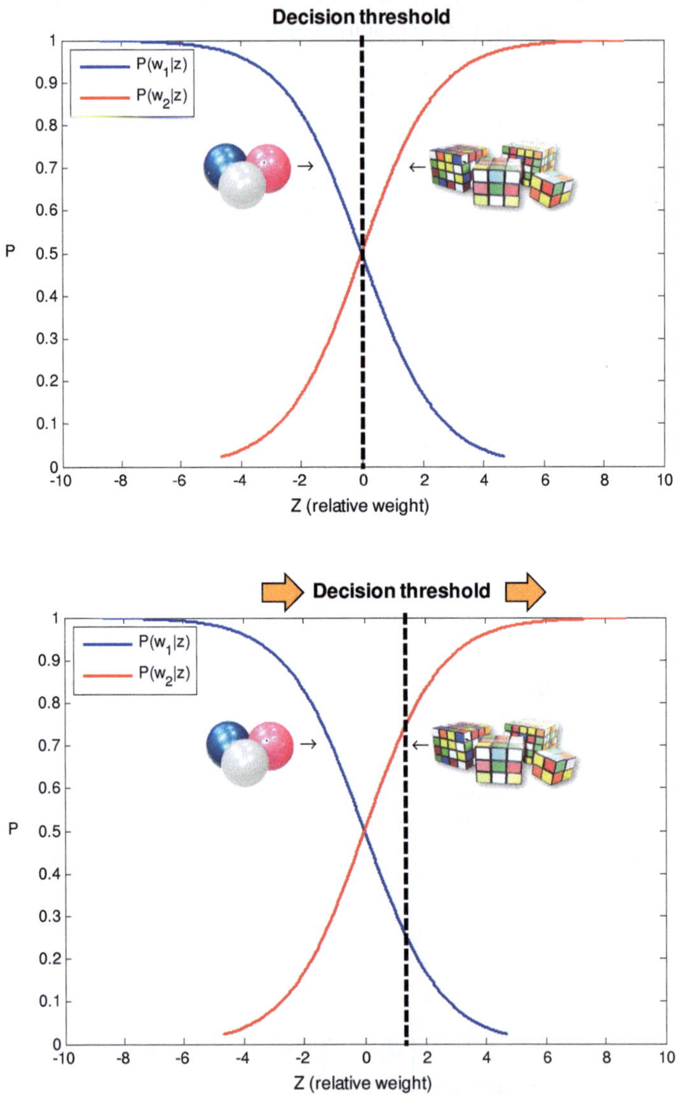

Fig. 5.5. Decision of a robotic perceptual system between a ball or a cube sampled from a pile by assigning risk. Top: example using the same sensor model, prior distribution and standard reference mass as in Fig. 5.4, and a zero-one loss function (see Tab. 5.1) — this means the robot will decide as if following a Maximum A Posteriori decision rule as in the top example of Fig. 5.4. Bottom: example with everything as before, except now with an assymetric loss function with $\Lambda(d_1 \mid w_2) > \Lambda(d_2 \mid w_1)$ — this means that the decision threshold is moved to the right (cf. with the bottom example of Fig. 5.4).

For example, if one uses the *zero-loss function* given by

$$\Lambda(a_i \mid s_j) = \begin{cases} 0, & i = j, \\ 1, & i \neq j, \end{cases} \quad \text{with } i, j = 1, 2, \ldots, N. \tag{5.4}$$

for a discrete state space (see Table 5.1 for an example)[4], then we find that $R(a_i \mid s_j) = 1 - P(s_j \mid O)$, and therefore minimising conditional risk corresponds to maximising the posterior (i.e., finding its *mode*). Therefore, the decision rule in this case is based on the MAP point estimate, and is by far the most common decision rule used in practical applications.

An alternative would be the least-squares estimator, also called minimal variance (MV) estimator or *square loss function*, which corresponds to the loss function given by $\Lambda(a_i \mid s_j) = (s_j - a_i)^2$. For this function, decisions which assume state estimates which are close, but not identical, to the real state are rewarded. In this case, as shown in section 1.2.4, the optimal decision is simply the *mean* of the posterior distribution. Another relatively common loss function would be the *linear or absolute loss function* given by $\Lambda(a_i \mid s_j) = \|s_j - a_i\|$, which can similarly be shown to correspond to the optimal decision given by the *median* of the posterior distribution. These alternatives to the MAP-based decision rule become really useful only if the posterior distribution is multimodal, in which case they will obviously tend to be more robust; if not (e.g., if the distribution is normal), all of these loss functions yield the exact same estimate, given that the mode, the mean and the median of the posterior will refer to the exact same value.

5.3 Dynamic Bayesian Decision

We will now consider the case where the decision made at time t, not only depends on the current observation and prior knowledge of the state, but also on previous decisions – this is called *dynamic Bayesian decision*.

5.3.1 Decision-Theoretic Planning – Markov Decision Processes (MDP) and the Efferent Copy

Decision-theoretic planning refers to the practical application of BDT in a dynamical context to sequential decision problems.

In this context, let us start by assuming that our robot has factored out any conditional dependence of the perceptual state on the observations – formally, in this case we say that the state is *fully observable*. If we further

[4] Or equivalently the loss function given by $\Lambda(a_i \mid s_j) = -\delta(s_j - a_i)$, where δ is the Dirac delta function, for a continuous state space.

assume that the decision process follows the Markov assumption, our model of this process becomes

$$P({}^{0:T-1}A \wedge {}^{0:T}S) = P({}^{0}A \wedge {}^{0}S) \prod_{t=1}^{T} \left[P({}^{t-1}A) P({}^{t}S \mid {}^{t-1}A \wedge {}^{t-1}S) \right] \quad (5.5)$$

Applying a payoff, gain or reward function r_t to this model will associate states and actions with a number that quantifies their reward, $r_t : {}^{t}s \times {}^{t}a \mapsto \mathbb{R}$. This is a deterministic function that defines the robot's current goal, therefore allowing it to decide on future actions (again, refer to Fig. 5.1 for an overview of the big picture). A simple payoff function could be

$$r_t({}^{t}s, {}^{t}a) = \begin{cases} +50, & \text{if action } {}^{t}a \text{ leads to a goal state } {}^{t}s, \\ -5, & \text{otherwise.} \end{cases}$$

This function "rewards" the robot with a score of +50 if a goal state is reached, and "penalises" it with a score of −5, otherwise[5].

The robot designer's objective in this context is to devise a system that generates actions so as to optimise future expectations on payoff; an appropriate approach would be to choose actions so that the expectation of the sum of all future payoff[6], the *expected cumulative payoff*, given by

$$R_T = E\left[\sum_{\tau=1}^{T} \gamma^{\tau} r_{t+\tau} \right], \quad (5.6)$$

is maximised. Note that the expectation is computed over future momentary values for payoff $r_{t+\tau}$ earned by the robot from t to $t+\tau$, and that each payoff is weighted by the so-called *discount factor*, denoted as γ^{τ} (with $0 \leq \gamma \leq 1$). The effect of the discount factor is analogous to the variation of the value of currency – applying this analogy, $\gamma < 1$ would correspond to inflation (earlier payoffs are more important than later payoffs), while if $\gamma = 1$, whatever the temporal offset τ, all future payoffs would have the same weight.

The (probability theory-related) model defined by equation (5.5) within the context of the Markov assumption, together with the (utility theory-related) expected cumulative payoff given by equation (5.6), describe the most ubiquitous process in dynamic Bayesian decision, the fully observable *Markov decision process (MDP)*.

Given that R_T is a sum taken over T instants in time, T is called the *planning horizon*. Depending on this horizon, dynamic Bayesian decision processes are classified into three different cases:

[5] Note that this could be the precursor of an emotional background – an inkling, perhaps, of a pleasure/pain system – for an artificial cognitive system!

[6] An alternative example, not analysed here, is maximising the average of future payoff instead.

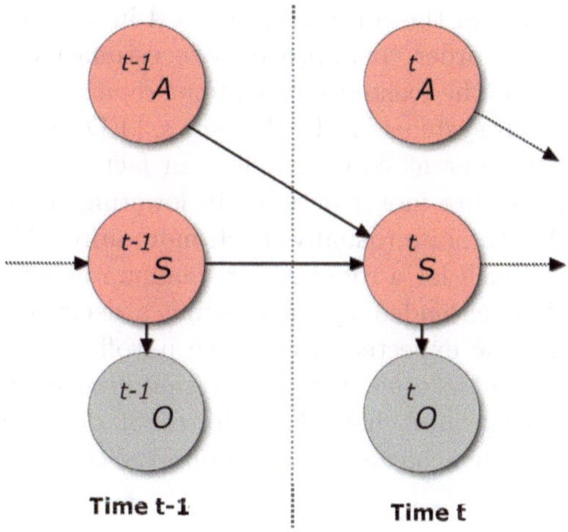

Fig. 5.6. Bayes network of the input-output hidden Markov model, the core model of partially observable Markov decision processes.

The greedy case. This corresponds to the case where $T = 1$, where the robot only attempts to maximise the single following payoff, and are known to be computable in polynomial time [7]. Note that, although the actual value of the discount factor is arbitrary in this case, it must be greater than zero.

The finite-horizon case. This corresponds to the case where $T \in (1, \infty)$. Typically, the discount factor is taken as equal to one; however, to compensate for the differences which commonly set apart the beginning and the end of a sequence of actions, different plans must be maintained, thus adding undesired complexity to the problem at hand.

The infinite-horizon case. In this case, the discount factor *must* be less than one, so as to make the expected cumulative payoff converge, assuming that the payoff is bounded by $r_{\max}$, to a finite value: $R_\infty \leq \frac{r_{\max}}{1-\gamma}$. However, it avoids the downsides of the finite-horizon case.

Now, imagine that we wish to integrate observations into our probabilistic model – after all, our robot will realistically only be able to partially observe the surrounding world at any single instant in time. In this case, equation (5.5) is further detailed by merging it with the DBN first-order Markov model, therefore restating it as

$$P({}^{0:T-1}A \wedge {}^{0:T}S \wedge {}^{0:T}O) = \\ P({}^{0}A \wedge {}^{0}S \wedge {}^{0}O) \prod_{t=1}^{T} \left[P({}^{t-1}A) P({}^{t}S \mid {}^{t-1}A \wedge {}^{t-1}S) P({}^{t}O \mid {}^{t}S) \right] . \tag{5.7}$$

If the state space is discrete, this model is also commonly named an *input-output hidden Markov model* – see Fig. 5.6.

"Reading" the values of the control variables A in such a model after they have been decided, in order to improve state estimation or prediction, is one possible reason for the existence of *efferent copies* of motor variables in animal central nervous systems [3]. In Wolpert's TED talk, he demonstrates the usefulness of the efferent copy, which is in fact an internal simulation of one's prospective future motor action[7], in lowering the unreliability of a motor command due to noise (usually much more unreliable than we would wish to imagine, as much for a robot as for a human being).

If equation (5.7) is extended by introducing a decision process that attempts to maximise the expected cumulative payoff, we are in presence of the so-called *partially observable Markov decision process (POMDP).*

Given all this background, how does the robot designer implement his objective in practice? He achieves this by proposing *control policies*, denoted as

$$\mathrm{Pol} : {}^{0:T-1}A, {}^{0:T}O \mapsto {}^{T}A, \tag{5.8}$$

which, in the case of full observability, reduces to

$$\mathrm{Pol} : {}^{T}S \mapsto {}^{T}A. \tag{5.9}$$

In summary, a control policy is a function that maps past observations into control actions, or states into control actions, when the state is fully observable. If the control policy only relies on the most recent observations and decisions, it will be purely *reactive* and fast, therefore allowing for real-time implementation; as one starts augmenting the memory required for the policy, we increase the degree of elaboration of the planning process at the expense of computational and temporal resources.

Practical implementation is then a question of finding the *optimal control policies* for each case, i.e. the policy that best fulfils the robot's goal by ensuring the highest expected discounted future payoff (note that fully observable processes are much more tractable in this respect, and hence their enduring popularity).

When the planning process does lead to intractable computations, it is often approximated by a type of iterative algorithms denominated *policy value iteration*. These algorithms start with random policies and improve them until some convergence criterion is met. An extensive amount of research has been and is still being carried out in this respect – the reader is advised to refer to [7] for a comprehensive description of such methods. An alternative is to try to automatically find hierarchical decompositions of the state space, so as to mitigate the combinatorial explosion – see, for example, the work by

---

7 Have you ever watched some intense physical activity (e.g. a sport) on TV, and a part of your body involuntarily moves so as to perform the action you are watching? That would be the result of a reflexive embodiment of an efferent copy.

Hauskrecht, Meuleau, Kaelbling, Dean, and Boutilier [16] or more recently by Pineau and Thrun [13].

5.3.2 Probabilistic Mapping and Localisation

In most applications, to be useful, autonomous mobile robots need to know *where* they are in the world. The process of a robot finding out its pose (i.e. position and bearing) within the map it is keeping of the environment (i.e. its internal representation of the surrounding world – see Chapter 2), as with any other autonomously moving object or being, is called *localisation*.

Localisation methods have been classified as *local or tracking techniques*, when the robot approximately knows its initial location and then keeps compensating for odometry errors as it moves, and *global techniques*, designed to estimate the position of the robot even under global uncertainty, when the robot does not know its initial location. The former typically cannot recover if they lose track of the robot's position (within certain bounds), while the latter, albeit much more complex, solve the *wake-up robot problem* (or its "unconscious version", the *kidnapped robot problem*, where the robot is carried to an arbitrary location *during* it's operation), in that they can localise a robot without any prior knowledge about its position. In summary, global localisation methods are more powerful than tracking techniques, since they can typically deal with situations in which the robot is likely to experience serious positioning errors.

The most well-known global localisation technique would arguably be *Markov localisation* – the following text and example closely follows the excellent explanation given by Fox, Burgard, and Thrun [15].

Markov localization is a probabilistic model that assumes that the environment is *static* (i.e. the Markov assumption), and instead of maintaining a single hypothesis as to where in the world a robot might be, it maintains a probability distribution over the space of all such hypotheses. The probabilistic representation allows it to weight these different hypotheses in a mathematically sound way. Markov localisation is a specialised version of the IO-HMM presented in the previous section.

Example 5.5. **Simple example of Markov localisation**

Consider the situation depicted on Fig. 5.7. For the sake of simplicity, we assume that the space of robot positions is unidimensional – the robot can only move horizontally and it may not rotate. Now suppose the robot is placed somewhere in the environment depicted on the picture, of which it maintains a map, but it is not told its location. Markov localization represents this state of uncertainty by a uniform distribution over all positions, as shown by the graph in the case represented at the top of Fig. 5.7. At a given moment, the robot uses its sensors to find out that it is next to a door.

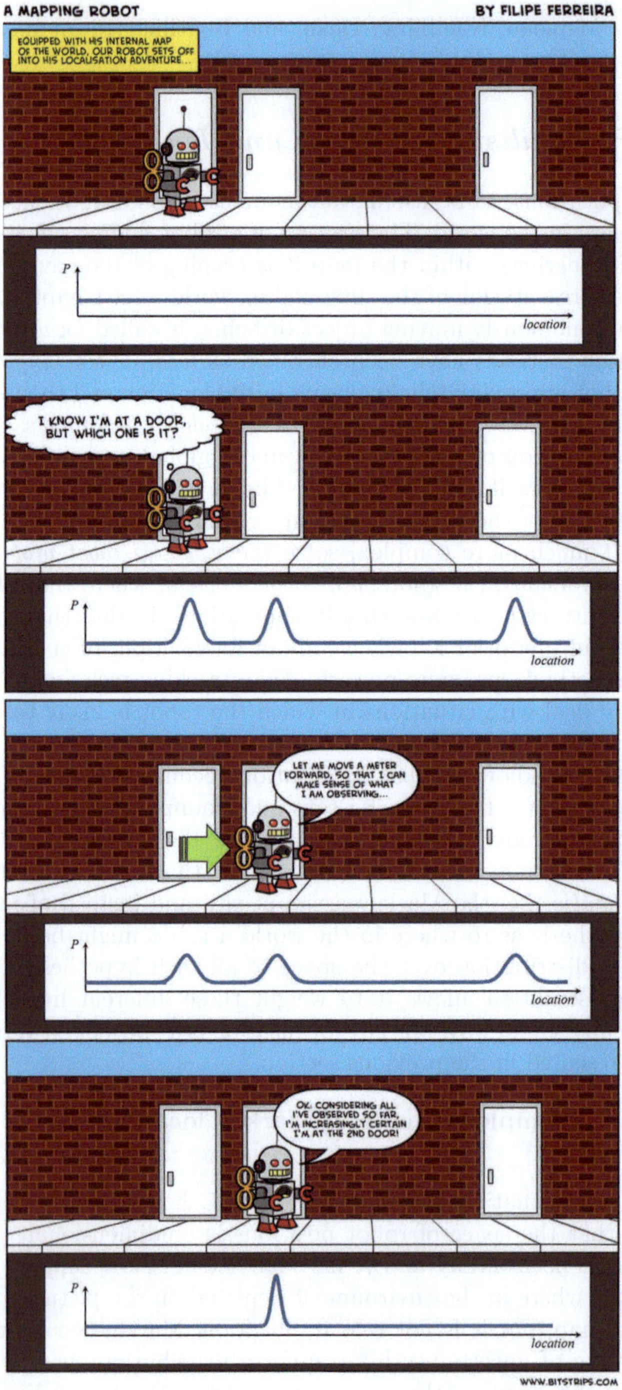

Fig. 5.7. Markov localisation – toy example, adapted from [15]. Refer to main text for details.

Markov localization updates the posterior on the robot's location by raising the probability for places next to doors, and lowering it anywhere else. This is illustrated in the second case depicted in Fig. 5.7. Note that the resulting probability distribution is multimodal (i.e. it has *multiple modes* appearing as local maxima), reflecting the fact that the available information is insufficient for unequivocal global localisation (although, at any given instant, the robot tries to guess its most plausible location). Also note that places not next to a door still possess non-zero probability. This is due to the inherent perceptual uncertainty: a single observation of a door is typically insufficient to exclude the possibility of not being next to a door.

Now let us assume the robot moves a meter forward. Markov localization incorporates this information by shifting the posterior distribution accordingly, as can be seen in the third case in Fig. 5.7. To account for the inherent uncertainty caused by robot motion, the new posterior is therefore "smoother" than the previous one. Finally, let us assume the robot performs a second observation using its sensors, and again finds itself next to a door. Now this observation is multiplied by the current (non-uniform) distribution, which leads to the final posterior used for estimation shown on the last case in Fig. 5.7. At this instant, most of the probability is centred around a single location, and therefore the robot is now quite certain about its position.

Concluding that typical approaches, such as Markov localisation or variations of Kalman filters including action variables, were not satisfactory enough, given the inherent separation of localisation and control models, Diard and Bessière [4] used an abstracted hierarchy, introduced in Chapter 4, to define the so-called *Bayesian maps*, a generalisation of Markov localisation.

They defined several Bayesian maps corresponding to various locations in the environment, each of which a model of sensorimotor interactions with a part of the environment. Then, they built an *abstracted map* based on these models. In this new map, the location of the robot is defined in terms of the submap that best fits the observations obtained from the robot's sensors. The main goal of their abstracted map was to navigate in the environment; therefore, they were more interested in the action to be performed than the actual location of the robot. Nevertheless, the action chosen by their model at each instant is made with respect to the uncertainty of the location, and localisation itself is an optional goal, to be used in a situation where human-robot or robot-robot communication would be of interest, for example.

5.4 Attention- and Behaviour-Based Action Selection

Instead of trying to find a structural decomposition for the state space automatically, an alternative approach would be to incorporate knowledge about

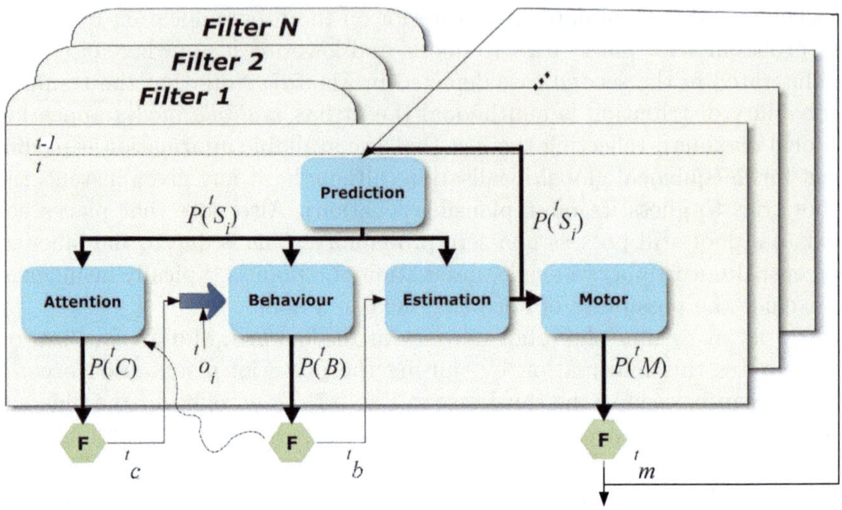

Fig. 5.8. Depiction of global and local processes relating to the global filter and the elementary filters in the attention- and behaviour-based action selection model by Koike et al. [5]. Please refer to main text for an explanation of coherence-based fusion, denoted here as an "F" enclosed inside an hexagon. The hatched arrow indicates the behaviour probability distribution is predicted at the moment the attention question is dealt with.

the goals, tasks or domain directly in the model – no utility would therefore be assigned, and a MAP-based decision method could be applied directly to the posterior on actions and/or states.

Koike, Bessière, and Mazer [5] proposed to separate the global filter model into i *elementary filters*, so as to reduce the complexity and dimensionality of the local state space partitions arising from the independence condition between state ${}^tS_i$ and observation variables ${}^tO_i$. Additionally, Koike et al. replaced action variables by *explicit motor variables* ${}^tM$ in the model.

Attention has been a widely studied process in neuroscience and psychophysics – it is known to be the process by which the brain sends out commands to redirect the sensors (in which case, it is classified as *overt attention*, and is a part an overall *active perception* process), and reassigns computational resources for perception (in which case, it is classified as *covert attention*), in both cases to process specific sensory features. On the other hand, a *behaviour* can be defined as a pattern of actions that can be observed in an agent implementing a given task to fulfil its goals, according to its own perception of the world. As such, and considering our introductory section, we can adapt this notion so as to propose that a behaviour prescribes a *collection of motor patterns*.

Program
Description
Specification

Relevant variables (sequences taken from $t = 0 \ldots T$):

${}^{0:T}S_i$: sequence of state variables for elementary filter i;

${}^{0:T}O_i$: sequence of observation variables for elementary filter i;

${}^{0:T}C, {}^{0:T}\alpha_i$: sequence of global attention variables and respective coherence variable;

${}^{0:T}B, {}^{0:T}\beta_i$: sequence of global behaviour variables and respective coherence variable;

${}^{0:T}M, {}^{0:T}\lambda_i$: sequence of global motor variables and respective coherence variable.

Decomposition:

$$P({}^{0:T}S_i \wedge {}^{0:T}O_i \wedge {}^{0:T}C \wedge {}^{0:T}\alpha_i \wedge {}^{0:T}B \wedge {}^{0:T}\beta_i \wedge {}^{0:T}M \wedge {}^{0:T}\lambda_i \mid \pi_i) =$$

$$P({}^{0}S_i \wedge {}^{0}O_i \wedge {}^{0}C \wedge {}^{0}\alpha_i \wedge {}^{0}B \wedge {}^{0}\beta_i \wedge {}^{0}M \wedge {}^{0}\lambda_i \mid \pi_i)$$

$$\prod_{t=1}^{T} \begin{bmatrix} P({}^{t}S_i \mid {}^{t-1}S_i \wedge {}^{t-1}M \wedge \pi_i) \\ \times P({}^{t}O_i \mid {}^{t}S_i \wedge {}^{t}C \wedge \pi_i) \\ \times P({}^{t}C \mid \pi_i) \times P({}^{t}\alpha_i \mid {}^{t}C \wedge {}^{t}B \wedge {}^{t}S_i \wedge \pi_i) \\ \times P({}^{t}B \mid \pi_i) \times P({}^{t}\beta_i \mid {}^{t}B \wedge {}^{t}S_i \wedge {}^{t-1}B \wedge \pi_i) \\ \times P({}^{t}M \mid \pi_i) \times P({}^{t}\lambda_i \mid {}^{t}M \wedge {}^{t}S_i \wedge {}^{t}B \wedge {}^{t-1}M \wedge \pi_i) \end{bmatrix}$$

Parametric forms:

$P({}^{0}S_i \wedge {}^{0}O_i \wedge {}^{0}C \wedge {}^{0}\alpha_i \wedge {}^{0}B \wedge {}^{0}\beta_i \wedge {}^{0}M \wedge {}^{0}\lambda_i \mid \pi_i)$: initial conditions;

$P({}^{t}S_i \mid {}^{t-1}S_i \wedge {}^{t-1}M \wedge \pi_i)$: local dynamic model for elementary filter;

$P({}^{t}O_i \mid {}^{t}S_i \wedge {}^{t}C \wedge \pi_i)$: local observation model for elementary filter;

$P({}^{t}C \mid \pi_i)$: prior on attention variables;

$P({}^{t}\alpha_i \mid {}^{t}C \wedge {}^{t}B \wedge {}^{t}S_i \wedge \pi_i)$: attention model in fusion with coherence form.

$P({}^{t}B \mid \pi_i)$: prior on behaviour variables;

$P({}^{t}\beta_i \mid {}^{t}B \wedge {}^{t}S_i \wedge {}^{t-1}B \wedge \pi_i)$: behavioural model in fusion with coherence form;

$P({}^{t}M \mid \pi_i)$: prior on motor variables;

$P({}^{t}\lambda_i \mid {}^{t}M \wedge {}^{t}S_i \wedge {}^{t}B \wedge {}^{t-1}M \wedge \pi_i)$: motor model in fusion with coherence form.

Identification:

Defined a priori by designer or through learning method (Chapter 6).

Questions:

$P({}^{t}S_i \mid {}^{0:t-1}o_i \wedge {}^{0:t-1}m \wedge {}^{0:t-1}c \wedge [{}^{0:t-1}\alpha = 1] \wedge [{}^{0:t-1}\beta = 1] \wedge [{}^{0:t-1}\lambda = 1] \wedge \pi_i)$: state prediction;

$P({}^{t}C \mid {}^{0:t-1}o_i \wedge {}^{0:t-1}m \wedge {}^{0:t-1}c \wedge [{}^{0:t}\alpha = 1] \wedge [{}^{0:t-1}\beta = 1] \wedge [{}^{0:t-1}\lambda = 1] \wedge \pi_i)$: attention selection;

$P({}^{t}B \mid {}^{0:t}o_i \wedge {}^{0:t-1}m \wedge {}^{0:t}c \wedge [{}^{0:t}\alpha = 1] \wedge [{}^{0:t}\beta = 1] \wedge [{}^{0:t-1}\lambda = 1] \wedge \pi_i)$: behaviour selection;

$P({}^{t}S_i \mid {}^{0:t}o_i \wedge {}^{0:t-1}m \wedge {}^{0:t}c \wedge [{}^{0:t}\alpha = 1] \wedge [{}^{0:t}\beta = 1] \wedge [{}^{0:t-1}\lambda = 1] \wedge \pi_i)$: state estimation;

$P({}^{t}M \mid {}^{0:t}o_i \wedge {}^{0:t-1}m \wedge {}^{0:t}b \wedge {}^{0:t}c \wedge [{}^{0:t}\alpha = 1] \wedge [{}^{0:t}\beta = 1] \wedge [{}^{0:t}\lambda = 1] \wedge \pi_i)$: motor commands.

Fig. 5.9. Bayesian program for the elementary filters π_i in the attention- and behaviour-based action selection model by Koike et al. [5].

Consequently, partitioning is accomplished by Koike et al. through the intermediation of *attention selection* on state-to-observation conditioning through global variable ${}^tC$, and of *behaviour selection* on state-to-motor conditioning through global variable ${}^tB$, therefore assembling collections of relevant environment features and motor patterns, respectively. The model is described by the diagram and corresponding Bayesian program presented in Fig. 5.8 and Fig. 5.9, respectively.

While conditional independence can be used to perform fusion for state and observation models, as explained in Chapter 2, this is not possible for the attention and behaviour models – however, this fusion is absolutely necessary, since these models relate each elementary filter to the global variables C and B, therefore creating a problem in the definition of the joint distribution equation. This is due to a definition of a model on a global variable cannot be performed as a product such as, in the case of a motor model,

$$\prod_{i=1}^{N_i} P({}^tM \mid {}^t S_i \wedge {}^{t-1} M \wedge \pi_i).$$

The reason for this is that ${}^tM$, as described in Chapter 3, cannot appear more than once on the left therefore creating invalid directed loops. Koike et al. [5] circumvented this problem by resorting to intermediate binary variables called *coherence variables*, such as ${}^t\lambda_i$ in the case of the motor model, thereby resulting in valid reformulations of these models, such as

$$P({}^tM \mid \pi_i) \prod_{i=1}^{N_i} P({}^t\lambda_i \mid {}^t M \wedge {}^t S_i \wedge {}^{t-1} M \wedge \pi_i).$$

Expressing a model in this way leads to what Pradalier, Colas, and Bessiere [12] called *coherence-based fusion*, and it does not add new knowledge or new assumptions, since the coherence variables are *always instantiated as being true* (i.e., global models are always assumed coherent).

Experiments with this framework were performed in simulation and with the indoor robot shown in Fig. 5.10, within the context of the European project BIBA (IST-2001-32115). The robot would navigate in an office-like environment, avoiding obstacles. A situation simulating the presence of a "predator" (to avoid), a "prey" (to chase) and a "home" (to return to) was enacted. Whenever the robot would perceive a predator, it would remain motionless if the predator was far away, or it would escape in the opposite direction if the predator was close. When a prey was seen, the robot would chase it, and when the prey was close enough, the robot would "capture" it[8].

In spite of the number of variables involved and the joint distribution's size used for these experiments (fifty four variables and thirty six terms, respectively), the programming task was simplified by the independence of the

[8] The authors provided a video of the robot's behaviours at `http://www.bayesian-programming.org/videoB1Ch8-1.html`.

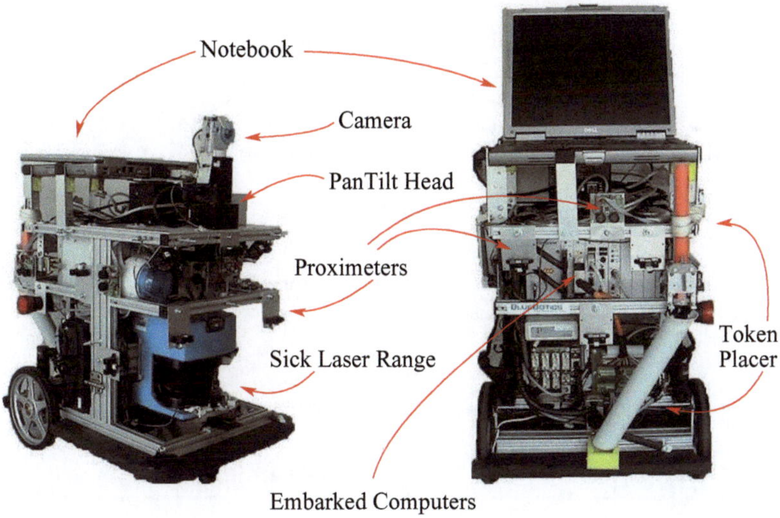

Fig. 5.10. The BIBA robot (reproduced by kind permission from [5]). This robot was devised and used within the context of the European project BIBA (IST-2001-32115).

elementary filters. The requirements to have dissociated elementary filters are hard to meet, but once these conditions are fulfilled, the consequent independence for programming each filter is proven to attain the goals described in the beginning of this section.

5.5 An Example of Probabilistic Decision and Control

One of the most promising and challenging recent trends of research in the development of probabilistic models of the action-perception loop has been the investigation of the role of the efferent copy, a notion introduced earlier on in section 5.3.1. This approach will be potentially crucial in dealing with the stability of closed-loop systems when taking uncertainty into consideration.

An efferent copy is an internal copy of an outflowing (i.e., *efferent*) motor signal. As Wolpert [2] shows in his presentation, it is believed to be fed back with and compared with the (*reafferent*) sensory input that results from the effects of actuation, allowing simultaneously for processes similar to traditional closed-loop control, but also for the modulation of sensory input in order to remove the effects of self-motion. As mentioned in section 5.3.1, efferent copies enable cognitive systems to predict the effects of action.

In the following text, we will present an example that attempts to model the efferent copy in a probabilistic robotic action-perception loop.

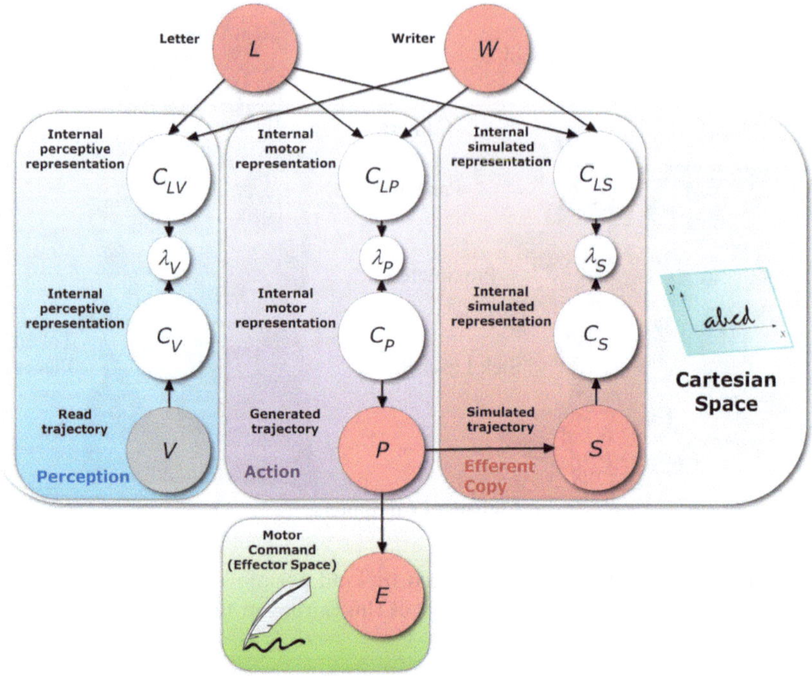

Fig. 5.11. Bayesian action-perception (BAP) framework for the study the interaction between the production and recognition of cursive letters proposed by Gilet et al. [1]

Example 5.6. **Closing the action-perception loop in the interaction between the production and recognition of cursive letters**

Gilet, Diard, and Bessière [1] proposed a complete mathematical formulation for the action-perception loop, which they called the Bayesian action-perception (BAP) model (see Fig. 5.11). The purpose of this model, according to the authors, is to study the interaction between the production and recognition of cursive letters, driven by the argument that the dual tasks of reading and writing have seldom been studied jointly.

The BAP framework includes a feedback loop emulating the efferent copy process. The authors' support the use of this loop with the fact that, in the study of handwriting, a growing body of literature has suggested a strong involvement of the motor system during the perception of letters. An illustrative example of such research can be found in the work by Longcamp et al. [11], in which a part of the motor cortex of the brain was found to be significantly activated during both reading and writing tasks. However, when the experiments involved visually presenting another class of stimuli – pseudoletters, which are as visually complex as letters, but for which the subjects had

Fig. 5.12. Robotic systems used as effectors for the BAP framework (reproduced from [1], with kind permission)

no previous experience in writing – the same motor area was not activated, thus strongly implying the involvement of efferent copies.

To control which part of the model – the perceptual, the action/actuation or the efferent copy branch – would be active at a given instant, Gilet et al. used a hierarchical combination of Bayesian switches, which we introduced in Chapter 4, one for each branch, as shown in Fig. 5.11. Another important feature of the framework which can be seen in the figure is the seamless transformation between Cartesian and effector space representations of space that is made possible by a probabilistic approach.

Gilet et al. showed how the BAP model solves the six cognitive tasks using Bayesian inference: i) letter recognition (purely sensory), ii) writer recognition, iii) letter production (with different effectors), iv) copying of trajectories, v) copying of letters, and vi) letter recognition (with internal simulation of movements) – see Fig. 5.12 for an overview of the robotic systems used in the experiments.

5.6 Final Remarks and Further Reading

An incredibly extensive amount of research has been amassed concerning Bayesian decision theory in the past few decades. For example, regarding computational models of perception and cognition, the reader is referred to the enthralling discussions presented by Ernst [6]; Ernst and Bülthoff [8]; Kersten, Mamassian, and Yuille [9] and in particular the seminal work by Yuille and Bülthoff [17].

Regarding decision-theoretic planning and MDPs and POMDPs, a good starting point would be the review by Boutilier, Dean, and Hanks [14], while a comprehensive study of probabilistic approaches to planning and control in terms of robotic applications is given by Thrun, Burgard, and Fox [7].

As in previous chapters, two important quick references to probabilistic approaches for decision can be found in [3; 10].

References

1. Gilet, E., Diard, J., Bessiére, P.: Bayesian Action–Perception Computational Model: Interaction of Production and Recognition of Cursive Letters. PLoS ONE 6(6), e20387 (2011), doi:10.1371/journal.pone.0020387 XXIII, 142, 143
2. Wolpert, D.: The real reason for brains. Video on TED.com (2011), `http://www.ted.com/talks/daniel_wolpert_the_real_reason_for_brains.html` 121, 122, 134, 141
3. Colas, F., Diard, J., Bessiére, P.: Common Bayesian Models For Common Cognitive Issues. Acta Biotheoretica 58(2-3), 191–216 (2010) 134, 144
4. Diard, J., Bessiére, P.: Bayesian maps: probabilistic and hierarchical models for mobile robot navigation. In: Bessiére, P., Laugier, C., Siegwart, R. (eds.) Probabilistic Reasoning and Decision Making in Sensory-motor Systems. STAR, vol. 46, pp. 153–176. Springer, Heidelberg (2008) 137
5. Koike, C.C., Bessiére, P., Mazer, E.: Bayesian Approach to Action Selection and Attention Focusing. In: Bessiére, P., Laugier, C., Siegwart, R. (eds.) Probabilistic Reasoning and Decision Making in Sensory-motor Systems. STAR, vol. 46, pp. 177–201. Springer, Heidelberg (2008) XXIII, 138, 139, 140, 141
6. Ernst, M.O.: A Bayesian view on multimodal cue integration. In: Human Body Perception From The Inside Out, ch. 6, pp. 105–131. Oxford University Press, New York (2006) 144

7. Thrun, S., Burgard, W., Fox, D.: Probabilistic robotics. MIT Press, Cambridge (2005) 133, 134, 144
8. Ernst, M.O., Bülthoff, H.H.: Merging the senses into a robust percept. Trends in Cognitive Sciences 8(4), 162–169 (2004) 123, 144
9. Kersten, D., Mamassian, P., Yuille, A.: Object perception as Bayesian inference. Annual Review of Psychology 55, 271–304 (2004) 144
10. Diard, J., Bessiere, P., Mazer, E.: A survey of probabilistic models using the Bayesian programming methodology as a unifying framework. In: International Conference on Computational Intelligence, Robotics and Autonomous Systems (IEEE-CIRAS), Singapore (2003) 144
11. Longcamp, M., Anton, J.L., Roth, M., Velay, J.L.: Visual presentation of single letters activates a premotor area involved in writing. Neuroimage 19(4), 1492–1500 (2003) 142
12. Pradalier, C., Colas, F., Bessiere, P.: Expressing Bayesian fusion as a product of distributions: Applications in robotics. In: Proceedings of the 2003 IEEE/RSJ International Conference on Intelligent Robots and Systems (IROS 2003), vol. 2, pp. 1851–1856 (2003) 140
13. Pineau, J., Thrun, S.: High-level robot behavior control using POMDPs. In: AAAI 2002 Workshop on Cognitive Robotics, vol. 107 (2002) 135
14. Boutilier, C., Dean, T., Hanks, S.: Decision-theoretic planning: Structural assumptions and computational leverage. Journal of Artificial Intelligence Research 11, 1–94 (1999) 144
15. Fox, D., Burgard, W., Thrun, S.: Markov Localization for Mobile Robots in Dynamic Environments. Journal of Artificial Intelligence Research 11, 391–427 (1999) 135, 136
16. Hauskrecht, M., Meuleau, N., Kaelbling, L.P., Dean, T., Boutilier, C.: Hierarchical solution of Markov decision processes using macro-actions. In: Proceedings of the Fourteenth Conference on Uncertainty in Artificial Intelligence, pp. 220–229 (1998) 135
17. Yuille, A.L., Bülthoff, H.H.: Bayesian decision theory and psychophysics. In: Knill, D., Richards, W. (eds.) Perception as Bayesian Inference, pp. 123–161. Cambridge University Press, Cambridge (1996) 144
18. Russell, S.J., Norvig, P., Canny, J.F., Malik, J.M., Edwards, D.D.: Artificial intelligence: a modern approach, 3rd edn. Prentice Hall, Englewood Cliffs (1995) 122

[illegible]

8. [illegible] M., [illegible] D.: Merging the [illegible] ... (2008) [illegible]

9. [illegible] D., [illegible] P., [illegible] (2011) [illegible]

10. [illegible]

11. [illegible]

12. [illegible]

[illegible]

6

Probabilistic Learning

Experience is the teacher of all things.
Commentarii de Bello Civili *(Commentaries on the Civil War), 2. 8,*
Julius Caesar (50s or 40s BC)

There can be no doubt that all our knowledge begins with experience.
Critique of Pure Reason, *B 1, Immanuel Kant (1781; 1787)*

Experience is a brutal teacher, but you learn. My God, do you learn.
Of Other Worlds: Essays and Stories, *Clive S. Lewis (2002)*

6.1 Introduction

An intuitive tell-tale of intelligence is the ability animals possess, particularly humans, of *learning from experience.* So, in fact, when we set out in designing *truly* intelligent systems in robotics, the general aim is to conjure up an architecture that is equally capable of:

- reasoning about the surrounding world given observed data, thereby generating a *representation* – see Chapter 2 to recall what this means in terms of perception;
- learning better representations for the future from the data it is gathering in the present, therefore preparing for *generalisation* – i.e., increasing cognitive performance by refining its internal model of the world as new data becomes available.

A subset of the artificial intelligence research area called *machine learning* is precisely about the construction of such artificial cognitive systems. Machine learning focusses on *prediction*, and the main goal of a learner is to improve its predictive capabilities from new data through generalisation based on what is called the *training data*, generically denoted as Δ. The training data is provided to the system with the hope of reflecting as closely as possible the nature of the model.

Machine learning algorithms are usually classified according either to the input they are provided with or to the desired outcome of their application, of which a partial taxonomy including the most commonly used types of algorithm is provided next:

J.F. Ferreira and J. Dias: *Probabilistic Approaches for Robotic Perception*,
Springer Tracts in Advanced Robotics 91,
DOI: 10.1007/978-3-319-02006-8_6,

Supervised learning – This is the simplest type of machine learning approach. These algorithms are externally guided during a *training phase* so as to generate a function that maps inputs to outputs (also called *labels*, because they are often provided by human experts labelling the training examples). For example, in a classification problem, the artificial learner approximates a function mapping a vector into classes by examining input-output examples of the function.

Unsupervised learning – In this case, the set of inputs is modelled automatically without labelling.

Semi-supervised learning – This method combines both the previous approaches in order to generate an appropriate function or classifier.

Reinforcement learning – In this case, the system learns how to act by observing the impact of each action on the environment, and using this feedback to define a set of rewards.

Probabilistic approaches, and more specifically Bayesian modelling, provide a unifying framework that is able to *explicitly and inherently* address perception, reasoning and learning. As a matter of fact, learning, in this perspective, is "just" another form of probabilistic decision – the cognitive system effectively decides on what its internal model should be considering the training data it is provided with. Let us elaborate further on this notion in the following section.

6.2 Probabilistic Learning as a Decision Process

Learning, from the probabilistic point-of-view, can manifest itself in one of two ways (or generically as a combination of both):

- through the estimation of the *parameters* defining the probability distributions that relate random variables in a model by previously assigning conditional plausibility, when the nature of the relations between variables (otherwise called the *structure* of the model, in an allusion to its graphical representation) is known or defined beforehand;
- through the determination of the structure of the model, when it is unknown.

Assume a generic probabilistic model we wish our robot to learn, represented by $\Pi : \Theta \wedge \Pi'$, where Π' represents the model's *parametric form or structure* and Θ represents the respective *parameters*, providing a domain governed by some underlying distribution P^*. As in section 4.3.3, we are given a set of i.i.d. samples of distribution P^* (i.e., the respective training data), denoted as $\Delta = \{\delta_1, \ldots, \delta_N\}$. We are also given a family of models (i.e., a collection of different structures with arbitrary parameters sharing the same variables in their joint distributions), and the task of our robot is to learn from the training data a model $\tilde{\Pi}$ in this family defining a distribution $P_{\tilde{\Pi}}$

(or simply $\tilde{P}$, when $\tilde{\Pi}$ is unambiguous). As understood from above, we may want the robot to learn only model parameters for a fixed structure, or a part or all of the model's structure. In some cases, we would wish the robot to be able to deal with a set of different hypotheses, storing within its artificial cognitive system, therefore, not a single model but rather a distribution over models, or alternatively an estimate of its confidence in the model that it learnt.

The ideal goal for our robot would be to conjure up a model $\tilde{\Pi}$ that would capture the distribution P^* from the sampled data Δ as precisely as possible. Unfortunately, due to the limited nature of the set of training data and also computational reasons, only a rough approximation of the underlying distribution will be obtainable, and hence our robot should be capable to literally cut its losses by defining what would be reasonable criteria for *deciding* which would be the *best* approximation to Π. Different models will, of course, in general embody different trade-offs: a model $\tilde{\Pi}$ may be better according to one metric, but worse according to another.

A logical course of action would be to formulate the robot's general learning goal as one of *density estimation*: constructing a model $\tilde{\Pi}$ such that $\tilde{P}$ is "close" to the generating distribution P^*. How do we evaluate the quality of this approximated model $\tilde{\Pi}$?

Let us first take a closer look at the nature of the training data. Fig. 6.1(a) shows a Bayesian network representing what could have been the classifier included in the cognitive system of the robot shown on Fig 1.1, with which we started Chapter 1. The training data used for learning its parameters is presented in Fig. 6.1(b). Each row of the data set, represented as δ_i, is called a *case* and represents a set of random variable instantiations composed of an array of observations (i.e., instantiations of the variables corresponding to the grey nodes in Fig. 6.1(a), $\{T, C, M, Y, G, R\}$) and respective collection of labels (i.e., instantiations of the variables corresponding to the classifying red nodes in the Bayesian network, $\{P, B, A\}$), taken during a supervised learning procedure. In some of the cases of this example, there is *missing data* (unavailable during the training phase, for some arbitrary reason), represented with a "?" – this particular data set is thereby said to be *incomplete*. If all values were available, on the other hand, the data would be unsurprisingly called *complete*.

Knowing that our training data is obtained by sampling the underlying probability distribution P^* and organising these samples as cases representing instantiation sets of the model variables, one typical option to determine the quality of the approximated model $\tilde{\Pi}$ is to use the Kullback-Leibler divergence defined in section 1.4.3:

$$D_{KL}(P^*||\tilde{P}) = E_{P^*(\Delta)}\left[\log_2 \frac{P^*(\delta)}{\tilde{P}(\delta)}\right] = \sum_{\Delta} P^*(\delta) \log_2 \frac{P^*(\delta)}{\tilde{P}(\delta)}. \tag{6.1}$$

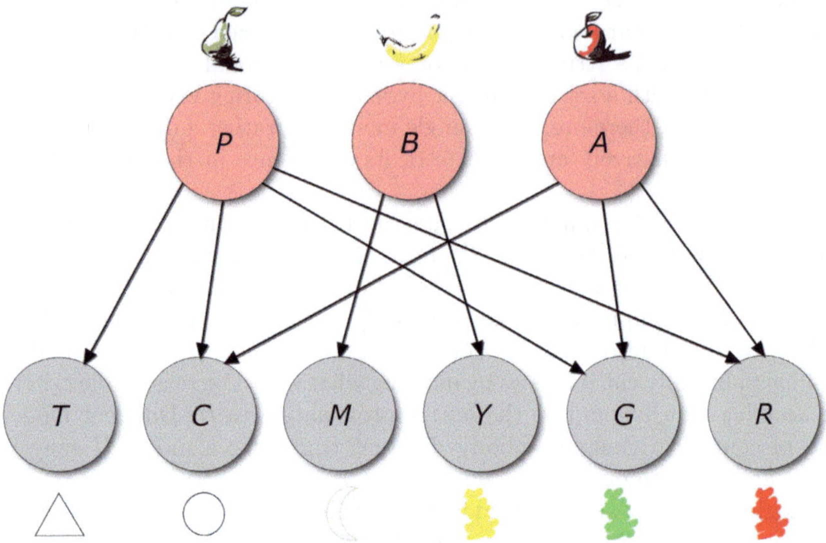

(a) Bayesian network for the classifier model.

Case	P	B	A	T	C	M	Y	G	R
1	T	?	F	T	F	F	F	T	?
2	F	?	T	F	T	?	F	F	T
3	F	T	F	F	F	T	T	F	F
4	T	F	?	T	T	F	?	T	T
⋮					⋮				

(b) Training data for the classifier model.

Fig. 6.1. Model and learning data for a (visual) fruit classifying robot. All random variables used in the model are binary (i.e., they all represent a *single proposition*, which is unequivocally either true or false): observed object shape and colour properties are represented by the grey nodes corresponding to variables T (triangular projection on image), C (circular projection on image), M (moon-shaped projection on image), Y (yellow), G (green) and R (red), respectively; object classes (labels) are represented by the red nodes corresponding to variables P (pear), B (banana) and A (apple).

As mentioned in section 1.4.3, this quality measure is zero when $\tilde{P} = P^*$ and otherwise positive, which means that the goal of our robot is to minimise this divergence.

To evaluate this metric as is it seems we would need to know $P^*(x)$, which in real-world applications is generally infeasible. However, this metric can be simplified as follows, in order to make it computable.

Proposition 6.1. *For any distributions P and P' over a set of variables $\mathcal{X}$ with respective set of instantiations Δ,*

$$D_{KL}(P||P') = -H_P(\mathcal{X}) - E_{P(\Delta)}\left[\log_2 P'(\delta)\right].$$

Proof.

$$\begin{aligned} D_{KL}(P||P') &= E_{P(\Delta)}\left[\log_2\left(\frac{P(\delta)}{P'(\delta)}\right)\right] \\ &= E_{P(\Delta)}\left[\log_2 P(\delta) - \log_2 P'(\delta)\right] \\ &= E_{P(\Delta)}\left[\log_2 P(\delta)\right] - E_{P(\Delta)}\left[\log_2 P'(\delta)\right] \\ &= -H_P(\mathcal{X}) - E_{P(\Delta)}\left[\log_2 P'(\delta)\right]. \end{aligned}$$

□

Applying this proposition to our specific problem by making $P = P^*$ and $P' = \tilde{P}$, we see that $-H_{P^*}(\mathcal{X})$, the negative entropy of P^*, does not depend on $\tilde{P}$; consequently, it does not influence the comparison between different approximate models. Therefore, our robot may focus on maximising the second term, $E_{P^*(\Delta)}\left[\log_2 \tilde{P}(\delta)\right]$, the *expected log-likelihood*, therefore devoting all its efforts to preferring models that allow the attainment of this goal. Note, however, that, although our robot is capable of comparing the relative quality of two models, using this rationale it is unable to establish the "absolute quality" of any chosen model (i.e., its quality comparing to the true model, the unknown optimum).

Until now, we have assumed that we wanted our robot to perform probabilistic inference using the learning model. However, this might turn out to be infeasible due to computational complexity, or even undesirable, since we might want our robot to focus on more specific conditional probability distributions and implicit relations between variables within a decomposition than the actual joint distribution. In this case, we may allow our robot to consider the learning framework as a decision process and apply a decision rule based on utility/risk assignments, exactly as described in Chapter 5, thus replacing the density estimation by point estimation, described in section 1.2.4. Moreover, this decision process may be considered as an ongoing process, much like what happens with humans, therefore enabling our robot to apply the dynamic decision methods described in the previous chapter, leading to *probabilistic reinforcement learning*.

In the following text, we will present methods for probabilistic parameter learning using complete and incomplete data, followed by probabilistic reinforcement learning, and finish by introducing the fundamentals of probabilistic structure learning.

6.3 Parameter Learning from Complete Data Using MLE

Assume that our robot observes several i.i.d. samples, denoted as $\Delta = \{\delta_1 \ldots \delta_N\}$, of a set of random variables $\mathcal{X}$ from an unknown distribution $P^*(\mathcal{X})$. Let us add to this assumption that the robot knows in advance the sample space it is dealing with (in other words, the composition of the set of random variables and the values each variable can take), but no further knowledge about the distribution $P^*(\mathcal{X})$ is available to it.

Next, we need to establish exactly what we want our robot to learn. Now, assume we have determined the *parametric model* of a particular distribution's parametric form for which the robot would have to estimate the free parameters (see section 3.2.3). Given a particular set of parameters $\Theta = \{\theta_1 \ldots \theta_M\}$ and an instance δ_i of $\mathcal{X}$, the parametric model $P(\delta_i \mid \Theta)$ assigns a probability to δ. Obviously, our robot needs that, for each choice of parameters Θ, $P(\delta_i \mid \Theta)$ is a legal distribution. In other words, it must be nonnegative and $\sum_{\Delta} P(\delta_i \mid \Theta) = 1$. In general, not all parameter values are allowed, and therefore our robot needs access to the definition of a *parameter space*.

Let us now define the likelihood function of Θ of a given parametric model as

$$L(\Theta \mid \Delta) \stackrel{\text{def}}{=} \prod_{i=1}^{N} P(\delta_i \mid \Theta). \tag{6.2}$$

Once our robot has this likelihood function, it can use it to choose the parameter values by applying the MLE decision rule,

$$\hat{\Theta} = \arg\max_{\Theta} L(\Theta \mid \Delta) = \arg\max_{\Theta} \prod_{i=1}^{N} P(\delta_i \mid \Theta). \tag{6.3}$$

The definition of the likelihood function given above can be extended so as to deal with conditional distributions, simply by repeating the MLE process for each possible instantiation of the conditioning variables, resulting in the so-called *conditional likelihood.*

But, even more importantly, the same process that is used to estimate the parameters of a single parameter model to maximise its likelihood can be used repeatedly for all distributions of a generative model's decomposition, with the ultimate desirable outcome of maximising the overall likelihood of the model. This is due to what is called the *global decomposability* of the likelihood function – for a detailed proof of this fundamental property, which is beyond the scope of this text, please refer to the excellent book by Koller and Friedman [6].

Let us now illustrate the MLE method with an example in which our robot attempts to learn a discrete Bayesian model from complete data.

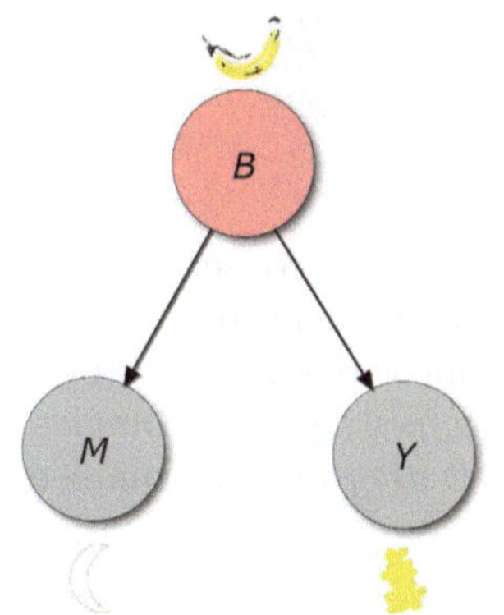

(a) Bayesian network for the detector model.

Case	B	M	Y
1	F	F	T
2	F	T	F
3	F	F	F
4	F	T	F
5	T	T	T
6	T	T	F
7	F	T	F
8	F	F	T
9	T	T	T
10	F	F	F
11	F	T	F
12	F	F	T
13	T	T	T
14	F	F	F
15	F	T	F
16	T	T	T

(b) Training cases of the training data Δ for the detector model.

B	M	Y	$P_\Delta(\bullet)$
F	F	F	$^3/_{16}$
F	F	T	$^3/_{16}$
F	T	F	$^5/_{16}$
F	T	T	$^0/_{16}$
T	F	F	$^0/_{16}$
T	F	T	$^0/_{16}$
T	T	F	$^1/_{16}$
T	T	T	$^4/_{16}$

(c) Empirical distribution for the detector model.

Fig. 6.2. Model and learning data for a (visual) robotic banana detector. All entities for (a) and (b) are defined as in Fig. 6.1; $P_\Delta(\bullet)$ is the so-called *empirical distribution*, a discrete data histogram that summarises the data set of b).

Example 6.1. **Training a simple object detector**

Consider Fig. 6.2 – the goal is to estimate the parameters of the Bayesian network from its training data set Δ using a supervised learning approach. We will assume that the sensors involved in generating the observations m_i and y_i in each case δ_i corresponding to the (correct) labels b_i (i.e., banana or non-banana) provided by the supervisor are precisely following the probabilities encoded in the actual Bayesian network, to which we do not have direct access.

Given this assumption, we can define a discrete data histogram, that implies that the empirical probability of a case with the specific instantiation $b \wedge m \wedge y$ is given by

$$P_\Delta(b \wedge m \wedge y) = \frac{\Delta\#(b \wedge m \wedge y)}{N},$$

where $\Delta\#(b \wedge m \wedge y)$ denotes the number of cases in data set Δ that are jointly instantiated by $b \wedge m \wedge y$, and N is the total number of cases in the data set.

This histogram can now be used to estimate the parameters of any of the conditional probabilities encoded by the model. Consider, for example, parameter $\theta_{[Y=\mathrm{T}]|[B=\mathrm{T}]}$, corresponding to the probability of a banana being sensed as being yellow $P([Y = \mathrm{T}] \mid [B = \mathrm{T}])$ (one of the two independent direct sensor models for the detector), is given by

$$P_\Delta([Y = \mathrm{T}] \mid [B = \mathrm{T}]) = \frac{P_\Delta([Y = \mathrm{T}] \wedge [B = \mathrm{T}])}{P_\Delta([B = \mathrm{T}])} = \frac{4/16}{5/16} = 4/5.$$

The discrete data histogram $P_\Delta(\bullet)$ that summarises the data set, given in Fig. 6.2(c) for the previous example, can now be formally defined.

Let Δ be a generic data set for a set of random variables given by a vector $\delta_1 \dots \delta_N$, where each δ_i represents a case consisting of a complete instantiation of those variables. The *empirical distribution* for this complete data set is defined as

$$P_\Delta(\alpha) = \frac{\Delta\#(\alpha)}{N}, \tag{6.4}$$

where $\Delta\#(\alpha)$ is the number of cases δ_i in the data set that satisfy the instantiation event α. Note that we have already shown in Example 1.6 that the empirical distribution corresponds to the binomial distribution's maximum likelihood estimate, and also for its generalised version, the multinomial distribution.

Given this definition, as in the previous example we propose that the process of estimating the parameter $\theta_{a|b}$ corresponding to the conditional probability $P(a \mid b)$ is defined as follows

$$\theta^{\mathrm{ML}}_{a|b} \stackrel{\mathrm{def}}{=} P_\Delta(a \mid b) = \frac{\Delta\#(a \wedge b)}{\Delta\#(b)}. \tag{6.5}$$

The count $\Delta\#(a \wedge b)$ is a *sufficient statistic* for multinomial distributions, since it contains all the information in the data set needed for the estimation task at hand [3; 6].

Example 6.2. **Training a simple object detector (continued)**

Lets get back to our robotic banana detector. Considering the network structure of Fig. 6.2(a), we are able to learn the parameters corresponding to each of the conditional probabilities involved – namely, the prior $P(B)$ and the likelihoods given by the sensor models $P(Y \mid B)$ and $P(M \mid B)$ – by repeatedly applying 6.5. The final outcome of the learning process can be seen in the CPTs presented in Table 6.1.

Let us now test a few inference conditions using our freshly trained model. For example, the probability of having perceived a banana given an object sensed as being yellow would be

Table 6.1. Final outcome of the supervised MLE learning process for the robotic banana detector of Fig. 6.2(a), in the form of conditional probability tables.

(a) Prior distribution (probability of an object being a banana).

B	θ_b^{ML}
F	$^{11}/_{16}$
T	$^{5}/_{16}$

(b) Likelihood of moon-shaped bananas (sensor model giving the probability of a banana yielding an observation of a moon-like shape).

B	M	$\theta_{m\mid b}^{\mathrm{ML}}$
F	F	$^{6}/_{11}$
F	T	$^{5}/_{11}$
T	F	0
T	T	1

(c) Likelihood of yellow bananas (sensor model giving the probability of a banana yielding an observation of a yellow object).

B	Y	$\theta_{y\mid b}^{\mathrm{ML}}$
F	F	$^{8}/_{11}$
F	T	$^{3}/_{11}$
T	F	$^{1}/_{5}$
T	T	$^{4}/_{5}$

$$
\begin{aligned}
P(b \mid y) &= \sum_M \frac{P(b)P(y \mid b)P(M \mid b)}{\sum_B P(B)P(y \mid B)P(M \mid B)} \\
&= \frac{P(b)P(y \mid b)P(m \mid b)}{\sum_B P(B)P(y \mid B)P(m \mid B)} + \frac{P(b)P(y \mid b)\overbrace{P(\neg m \mid b)}^{=0}}{\sum_B P(B)P(y \mid B)P(\neg m \mid B)} \\
&= \frac{P(b)P(y \mid b)\overbrace{P(m \mid b)}^{=1}}{P(b)P(y \mid b)\underbrace{P(m \mid b)}_{=1} + P(\neg b)P(y \mid \neg b)P(m \mid \neg b)} \\
&= \frac{^{5}/_{16} \times {^{4}/_{5}}}{^{5}/_{16} \times {^{4}/_{5}} + {^{11}/_{16}} \times {^{3}/_{11}} \times {^{5}/_{11}}} \\
&= 0.7458.
\end{aligned}
$$

What about the probability of having perceived a banana given an object sensed as *not* being moon-shaped? This would be given by

$$
\begin{aligned}
P(b \mid \neg m) &= \sum_Y \frac{P(b)P(Y \mid b)\overbrace{P(\neg m \mid b)}^{=0}}{\sum_B P(B)P(Y \mid B)P(\neg m \mid B)} \\
&= 0,
\end{aligned}
$$

which means that the MLE learning process has made it impossible for this model to recognise an object not shaped as a moon as ever being a banana, simply because the training phase never included the unlikely case (but still marginally possible, for anyone who knows his fruit) of a straight banana (no pun intended)!

(Stay tuned for the solution to this zero-probability problem later in this chapter...)

At this point, we can say that this is where the Bayesian approach meets traditional statistical inference, as will become clear in the following paragraphs.

A set of training data is no more than a sample of size N of cases defining the joint distribution of our model (as usual, assumed discrete). As the reader might have noticed, if different sets of training data with small number of cases are used, they will probably generate different values for the parameter estimates. However, as N tends toward infinity, (6.5) is asymptotically normal[1] [3].

This brings us to the following intuitive theorems, the proof of which is beyond the scope of this text[2].

Theorem 6.1.
Let Δ be a complete training data set. The parameter estimates defined in (6.5) *are the sole estimates that maximise the likelihood function*

$$\theta^{\mathrm{ML}} = \arg\max_{\theta} L(\theta \mid \Delta), \quad \text{iff } \theta^{\mathrm{ML}} = P_{\Delta}(a \mid b).$$

Theorem 6.2.
Let Δ be a complete training data set over model variables $\mathcal{X}$. Then

$$\arg\max_{\theta} L(\theta \mid \Delta) = \arg\min_{\theta} D_{KL}\left(P_{\Delta}(\mathcal{X}) || P_{\theta}(\mathcal{X})\right).$$

In summary, probabilistic learning using complete data may be performed using MLE, either by applying the empirical distribution method or by using the stock expressions for the MLE of the parameters of common conditional probability distributions on generic random variables (a fact which should shed a whole new light on the relevance of what was said in section 1.2.4 and Examples 1.6 and 1.7). The parameter estimates resulting from this technique have several important properties [3]: they are unique, they are asymptotically normal, they maximise the probability of the data, and, perhaps most importantly, they are easily computable by performing a single pass on Δ.

[1] As discussed in Chapter 1, this makes perfect sense according to the frequentist view.

[2] You can find the corresponding proofs in [3], if you are really eager to study them.

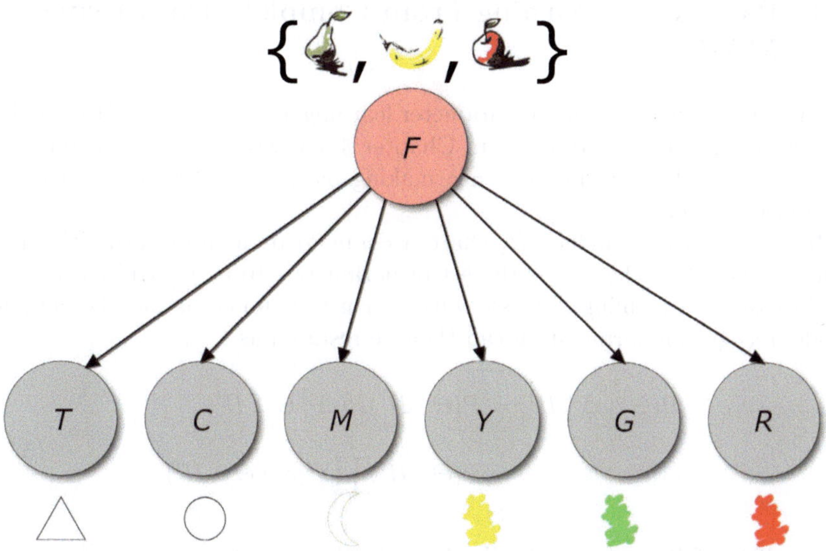

Fig. 6.3. Mutually exclusive version of the model and learning data for a (visual) fruit classifying robot (i.e., in this model, an object *is assigned to one, and only one, of the free available fruit classes*). Excepting the rather obvious fruit classifying random variable F, the remaining notation is as in Fig. 6.1.

We can therefore use this know-how in practice and generalise to models with random variables with arbitrary discrete support. We can either extend the empirical distribution method to discrete data histograms on n-ary model variables, or predefine standard parametric forms for the distributions used in the model, and in that case assume MLE parameter learning from the get-go. As an illustration of these options, the model represented in Fig 6.1 can be reformulated more compactly as in Fig. 6.3, with the added bonus of stating more rigorously that the propositions implied originally with B, P and A are mutually exclusive by introducing a single ternary fruit classifier variable including all of these classes, $F \in \{B, P, A\}$. One could then define the prior $P(F)$ (i.e., the probability of an object being either a banana, a pear or an apple, if nothing else is known), as, for example:

- a three-valued CDT, and use the empirical distribution method;
- having a standard parametric form instead – $P(F)$ could be assumed to be, for example, a discrete truncated Normal distribution, and a supervised learning process with complete data could then be used to estimate the respective mean and standard deviation parameters by applying the expressions derived in Example 1.7.

6.4 Parameter Learning From Complete Data Using MAP

Alternatively, we can think of parameter learning as an instance of the model recognition problem introduced in Chapter 4.3.3, where the parameters are the entities to be recognised, hence making parameter learning a Bayesian inference problem [2].

In this case, the models, Π, share a common parametric form, Π', and thus $\Pi : \Theta \wedge \Pi'$, where Θ is the set of parameters to be recognised. Let us again assume a training data set with N cases, denoted as Δ. The generic model recognition formulation can then be restated as

$$\begin{aligned} P(\Theta \wedge \Delta \mid \Pi') &= P(\Theta \mid \Pi')P(\Delta \mid \Theta \wedge \Pi') \\ &= P(\Theta \mid \Pi') \prod_{i=1}^{N} P(\delta_i \mid \Theta \wedge \Pi') \end{aligned} \tag{6.6}$$

where $P(\Theta \mid \Pi')$ is a prior distribution describing what is known beforehand about the parameters, and $P(\Delta \mid \Theta \wedge \Pi')$, or $P(\delta_i \mid \Theta \wedge \Pi')$ with the i.i.d. assumption, is the likelihood of the parameters.

So, learning can be accomplished by answering the following probabilistic query by applying a MAP decision process (making this a *Bayesian learning* process),

$$\begin{aligned} P(\Theta \mid \Delta \wedge \Pi') &\propto P(\Theta \mid \Pi')P(\Delta \mid \Theta \wedge \Pi') \\ &\propto P(\Theta \mid \Pi') \prod_{i=1}^{N} P(\delta_i \mid \Theta \wedge \Pi'). \end{aligned} \tag{6.7}$$

The likelihood functions are usually completely specified by the chosen parametric forms Π' and the parameters Θ of the model, so if the prior on the parameters $P(\Theta \mid \Pi')$ was to be assumed uniform, this would reduce to a MLE problem as in the previous section. However, if we wish the learning process to include an informative prior on the parameters as well, we would need additional parametrisation, and the so-called *hyperparameters* would have to be introduced, denoted here as Γ. The joint probability distribution would then become

$$P(\Gamma \wedge \Theta \wedge \Delta \mid \Pi') = P(\Gamma \wedge \Pi')P(\Theta \mid \Gamma \wedge \Pi')P(\Delta \mid \Theta \wedge \Pi'), \tag{6.8}$$

where $P(\Gamma \wedge \Pi')$ is a prior on the hyperparameters and $P(\Theta \mid \Gamma \wedge \Pi')$ is the distribution on the parameters conditioned on the hyperparameters.

As a result, the inference equation becomes

$$P(\Theta \mid \Delta \wedge \Pi') \propto \sum_{\Gamma} [P(\Gamma \wedge \Pi')P(\Theta \mid \Gamma \wedge \Pi')] \prod_{i=1}^{N} P(\delta_i \mid \Theta \wedge \Pi'). \quad (6.9)$$

This process of hyper-parametrisation can by repeated recursively, depending on what knowledge would the modeller wish to include in the model; however, it can be shown that deep layers of priors will have less and less effect on the parameters learning process as more layers are added to the learning hierarchy [2]. Note that, *if the parameters and respective priors are carefully chosen, Bayesian learning inherits the global separability property* of MLE[3].

MAP estimates can be computed in several ways:

- analytically, when the mode(s) of the posterior distribution can be given in closed form – this is the case when conjugate priors are used;
- using numerical methods.

Conjugate priors are $P(\Theta \mid \Gamma \wedge \Pi')$ distributions which are designed to be of *the same family* as the posterior distribution on the parameters. One can think of conditioning on conjugate priors as defining a kind of discrete time dynamical system: incoming data updates a set of hyperparameters through time, so one can take the variation in the hyperparameter set as a kind of chronological evolution of the system, corresponding to a dynamic learning process.

The usefulness of MAP estimation-based learning becomes clear when dealing with multinomial distributions for the likelihood on parameters with small training data sets. More specifically, imagine a situation where the random experiment constituting the training phase has been repeated an insufficient number of times, leading to a small number of trials and corresponding cases – if a particularly uncommon event is not represented in the dataset by at least one case, it will be unaccounted for, and thus irreparably assigned a null probability[4]. In other words, instead of just being assumed to be an unlikely event, it is deemed as impossible!

This situation is avoided by introducing a Dirichlet prior, the multivariate generalisation of the beta distribution and the conjugate prior of the multinomial distribution, that equals a uniform distribution within the range of the support allowed for each of the parameters of Θ, which is $0 < \theta < 1$. This would mean that we are absolutely ignorant regarding the parameters (see Chapter 1), apart from the fact that they are

[3] If the reader is interested in cases where this does not happen, for example when considering shared parameters, please refer to [6].

[4] Although similar, this is strictly different from having an incomplete data situation – there are no missing elements in each row of data, just insufficient rows/trials to have a statistically representative learning experiment.

neither 0 or lower, nor 1 or higher. This prior would be therefore defined as $P(\Theta \mid \Gamma \wedge \Pi') = \prod_i^N P(\theta_i \mid \Gamma \wedge \Pi') = 1$ for $0 < \theta_i < 1$, and 0 otherwise.

Given that the likelihood of $\theta = p$ given a single case $\delta_i = x_1 \wedge \ldots \wedge x_m$ is a multinomial distribution expressed as

$$L(p) = P(\delta_i \mid p \wedge \Pi') = \prod_{i=1}^{m} \left[p^{x_i} (1-p)^{1-p^{x_i}} \right] = p^s (1-p)^{m-s} ,$$

with s denoting the number of occurrences of the specific case δ_i, the posterior probability distribution resulting from Bayesian inference is therefore

$$\begin{aligned} P(p \mid \delta_i \wedge \Pi') &= \frac{p^s (1-p)^{m-s}}{\int_0^1 p^s (1-p)^{m-s} \, dp} \\ &= \frac{(m+1)!}{s! \, (m-s)!} p^s (1-p)^{m-s} . \end{aligned}$$

This is a beta distribution with expected value given by

$$\int_0^1 p \left[P(p \mid \delta_i \wedge \Pi') \right] dp = \frac{s+1}{m+2} .$$

This the so-called *rule of succession*, a formula introduced in the 18th century by Pierre-Simon Laplace. Since the conditional probability for an occurrence of $[a = T]$ in the next experiment, given the value of p, is just p, Kolmogorov's additivity axiom (Chapter 1) tells us that the probability of occurrence of $[a = T]$ in the next experiment is just the expected value of p. Therefore, the MAP estimate for the parameter corresponding to the probability of a given b is given simply by

$$\theta_{a|b}^{\mathrm{MAP}} = \frac{\Delta\#(a \wedge b) + 1}{\Delta\#(b) + 2} , \tag{6.10}$$

thus adding the so-called *"pseudo counts"* to the empirical counts, and thereby avoiding the zero-probability problem altogether.

Example 6.3. **Training a simple object detector (finalised)**

Let us return one final time to our robotic banana detector. Now resorting to "pseudo counts", we are able to relearn the parameters corresponding to each of the conditional probabilities involved, generating an outcome the CPTs presented in Table 6.2.

Let us again test the probability of having perceived a banana given an object sensed as *not* being moon-shaped, by computing

Table 6.2. Final outcome of the supervised MAP learning process for the robotic banana detector of Fig. 6.2(a), in the form of conditional probability tables

(a) Prior distribution (probability of an object being a banana).

B	θ_b^{ML}
F	$12/18$
T	$6/18$

(b) Likelihood of moon-shaped bananas (sensor model giving the probability of a banana yielding an observation of a moon-like shape).

B	M	$\theta_{m\mid b}^{\mathrm{ML}}$
F	F	$7/13$
F	T	$6/13$
T	F	$1/7$
T	T	$6/7$

(c) Likelihood of yellow bananas (sensor model giving the probability of a banana yielding an observation of a yellow object).

B	Y	$\theta_{y\mid b}^{\mathrm{ML}}$
F	F	$9/13$
F	T	$4/13$
T	F	$2/7$
T	T	$5/7$

$$P(b \mid \neg m) = \sum_Y \frac{P(b)P(Y \mid b)P(\neg m \mid b)}{\sum_B P(B)P(Y \mid B)P(\neg m \mid B)}$$

$$= \underbrace{\frac{P(b)P(y \mid b)P(\neg m \mid b)}{\sum_B P(B)P(y \mid B)P(\neg m \mid B)}}_{Term_y} + \underbrace{\frac{P(b)P(\neg y \mid b)P(\neg m \mid b)}{\sum_B P(B)P(\neg y \mid B)P(\neg m \mid B)}}_{Term_{\neg y}}.$$

Calculating the term related with condition y first, we have

$$Term_y = \frac{P(b)P(y \mid b)P(\neg m \mid b)}{\sum_B P(B)P(y \mid B)P(\neg m \mid B)}$$

$$= \frac{P(b)P(y \mid b)P(\neg m \mid b)}{P(b)P(y \mid b)P(\neg m \mid b) + P(\neg b)P(y \mid \neg b)P(\neg m \mid \neg b)}$$

$$= \frac{6/18 \times 5/7 \times 1/7}{6/18 \times 5/7 \times 1/7 + 12/18 \times 4/13 \times 7/13} = 0.2354,$$

while we can establish that the term related with condition $\neg y$ is given by

$$Term_{\neg y} = \frac{P(b)P(\neg y \mid b)P(\neg m \mid b)}{\sum_B P(B)P(\neg y \mid B)P(\neg m \mid B)}$$

$$= \frac{P(b)P(\neg y \mid b)P(\neg m \mid b)}{P(b)P(\neg y \mid b)P(\neg m \mid b) + P(\neg b)P(\neg y \mid \neg b)P(\neg m \mid \neg b)}$$

$$= \frac{6/18 \times 2/7 \times 1/7}{6/18 \times 2/7 \times 1/7 + 12/18 \times 9/13 \times 7/13} = 0.0519,$$

and, therefore,

$$P(b \mid \neg m) = Term_y + Term_{\neg y} = 0.2873,$$

which means that the zero-probability problem has been dealt with appropriately – probability is still low, but sensing an object as not moon-shaped does not overshadow everything else (namely, the substantial contribution of the object still being yellow, embodied by $Term_y$).

As a final remark, we would like to introduce a disclaimer: strictly speaking, since MAP is a point estimation process, it cannot be considered a *representative* Bayesian estimation method unless the parameter random variables θ_i are discrete. More rigorously, MAP estimation is a limiting case of Bayes estimation under a zero-one loss function. The actual Bayesian estimator is, in fact, the *full distribution* inferred from the parameter model; however, the so-called "proper" Bayesian point estimators are usually reported by the posterior mean or median of the posterior distribution instead, together with credible intervals, since these estimators are optimal under squared-error and linear-error loss, respectively. Nevertheless, in the context of parameter learning for robotic perception, where one is commonly interested in identifying point estimates for the respective models with minimum fuss, we find MAP-based learning much more pertinent to address herewith and leave the alternatives for the reader to study elsewhere.

6.5 Parameter Learning from Incomplete Data – The Expectation-Maximisation (EM) Algorithm

There are many different reasons for having incomplete data for learning to start with – some of the variables may simply be unobservable or *hidden*[5] (this would result in a full column of missing data), or some of the propositions encoded on their support might be difficult to ascertain in practice (which would result in missing data scattered throughout the data set, as in Fig. 6.1). For example, in perceptual terms, it might be difficult to register either the "non-sensation" of something or the sensation of "nothing", to be used as training data; however, there is a definite advantage in integrating knowledge over these events in a perceptual model. One very popular way of learning parameters from incomplete data is the so-called *expectation-maximisation (EM) algorithm*, first proposed as a general-purpose algorithm by Dempster et al. [13], which we will be describing in the following text.

[5] In the literature, the terms "hidden variables" and "latent variables" have often been used interchangeably; for clarity, we opted to differentiate these two notions – we assume the former are known variables which are for some reason unobservable, while assuming the latter as factors which are not explicitly accounted for.

Algorithm 6.1. Expectation-maximisation algorithm. In the pseudocode below, $\mathcal{X}_{\text{obs}}$ represents the set of observed data and $\mathcal{X}_{\text{miss}}$ the set of missing data, and therefore $\mathcal{X}_{\text{obs}} \cup \mathcal{X}_{\text{miss}} \equiv \Delta$.

Input:
  Π: model with parametric forms Π';
  Θ: parametrisation of model Π;
  Δ: training data set of size N;
Output: EM parameter estimates for model Π

$k \longleftarrow 1$;
initialise (e.g., with random values) Θ^0;
while $\left|\Theta^k - \Theta^{k-1}\right| < c$ **do**
  `// ** E-step **`
  compute
  $Q(\Theta \mid \Theta^{k-1}) = E_{\mathcal{X}_{\text{miss}} \mid \mathcal{X}_{\text{obs}} \wedge \Theta^{k-1} \wedge \Pi'} \left[\log P(\mathcal{X}_{\text{obs}} \wedge \mathcal{X}_{\text{miss}} \mid \Theta^{k-1} \wedge \Pi')\right]$;
  `// ** M-step **`
  compute $\Theta^k = \arg\max_{\Theta} Q(\Theta \mid \Theta^{k-1})$;
end

Expectation-maximisation is a local search method – it first completes the data set by inducing an empirical distribution generated from initial values for all parameters of the model, and then uses it to perform parameter learning as if having complete data. The new parameter set Θ^k is guaranteed to have no less likelihood than the initial parameters, so this process can be iteratively repeated until some convergence condition is met [3].

All of what is described above is done implicitly in practice – the actual EM pseudocode is presented in Algorithm 6.1. Note that it can be as easily applied assuming MLE as it is assuming MAP (Bayesian learning). Incomplete data-based learning local search methods such as EM are not guaranteed to converge to a global maximum, so it is common in practice to run the algorithm several times with different initial values and select the solution that maximises the likelihood/posterior probability.

Many different variants and implementations of the EM method exist; for example, the popular *Baum-Welch algorithm* [14], based on the forward-backward inference algorithm and used for learning the parameters of discrete probabilistic models and HMMs, is a special case of EM. A particularly interesting implementation of the Baum-Welch algorithm is the incremental version presented by Florez-Larrahondo [10]; this is an online implementation, as opposed to the original Baum-Welch algorithm, in which model parameters are reestimated after each new observation. The algorithm can reportedly handle 10 sensorimotor samples per second.

6.6 Reinforcement Parameter Learning – Exploration *vs.* Exploitation and the Markov Decision Process Formulation of Learning

The standard *reinforcement learning model* is a dynamic decision process that plays out in a setting such as presented in Fig. 5.1 in the previous chapter. In each interaction step, the learning agent receives as input some indication of the current state of the environment. The agent then chooses an action to generate as an output. The action changes the state of the environment, and the value of this state transition is communicated to the agent through a scalar reinforcement signal that rewards (or not) the agent for its actions. The learning agent's behaviour should choose actions that tend to increase the long-run sum of values of the reinforcement signal. It can learn to do this through time by systematic trial-and-error, guided by one of a wide variety of algorithms that have been proposed in the literature.

One major difference between reinforcement learning and supervised learning is that a learner in this context *must explicitly explore its environment.* In order to illustrate the difficulties inherent to *exploration*, we will describe a well-known simple case.

The simplest possible reinforcement-learning problem is known as the k-armed bandit problem, which has been the subject of a great deal of study in the statistics and applied mathematics literature [9; 11]. Imagine our robot, a platform with a gripper, is in a room with a collection of k slot machines (called "one-armed bandits" in colloquial English, because they were originally operated by a lever on the side of the machine – the arm – instead of a button). The robot is permitted a fixed number of pulls, h, with his gripper, and any arm may be pulled on each turn. The machines do not require a deposit to play; the only cost is in wasting a pull playing a suboptimal machine. When arm i is pulled, machine i pays the robot 1 or 0, according to some underlying probability parameter p_i, where payoffs are independent events and the parameters p_i are unknown. What should the robot's strategy be?

This problem illustrates the fundamental *exploitation/exploration trade-off*. The robot might believe that a particular arm has a fairly high payoff probability; should it choose that arm all the time (exploitation), or should it choose another one that it has less information about, but currently believes to be worse (exploration)? Answers to these questions depend on how long the robot is expected to play the game; the longer the game lasts, the worse the consequences of prematurely converging to a suboptimal arm, and the more the robot should explore. There is a wide variety of solutions to this problem – for an overview, please refer to [11].

The reinforcement learning process becomes more interesting when extended so as to encompass more than just an immediate reward – in this case, they are formulated as Markov decision processes, and therefore follow the implementation guidelines presented in Chapter 5.

6.7 Structure and Nonparametric Learning

According to Kemp and Tenenbaum [8], "algorithms for finding structure in data have become increasingly important both as tools for scientific data analysis and as models of human learning". Standard algorithms, of which a comprehensive study is presented in [6], however, until recently could only learn structures of a single form that had to be specified in advance. To overcome this, Kemp and Tenenbaum presented a computational model that learns structures of many different forms and that discovers which form is best for a given dataset, claiming that their approach brings structure learning methods closer to human abilities and suggesting that it may lead to a deeper computational understanding of cognitive development.

Kemp and Tenenbaum's approach is equivalent to applying the abstracted hierarchy presented in section 4.3.4 as equation (4.10) to classes of models and their parameters. In fact, in that equation, when X is replaced by Θ, the abstracted hierarchy yields the Bayesian Model Selection (BMS) method, and when the corresponding generative model is used to jointly compute the distribution over models and parameters through $P(\Pi \wedge \Theta \mid \Delta)$ it reflects the approach proposed by those authors [2].

Finally, beyond simple structure learning, a special kind of state-of-the-art hierarchical Bayesian frameworks, known as "infinite" or nonparametric hierarchical models, allow for structure learning without the need to know the number of variable classes in advance. These models put forward an unbounded amount of structure, but compensate by only actively engaging a finite amount of degrees of freedom for a given data set. An automatic Occam's razor embodied in the Bayesian inference process using these frameworks trades off model complexity and fit to ensure that new structure (i.e., a new class of variables) is introduced only when the data truly requires it [1].

6.8 Examples of Probabilistic Learning

We will now present a simple example of probabilistic learning – a continuation of an example from Chapter 2, followed by two brief references to illustrative examples that the reader might want to look up.

Example 6.4. **Learning the free parameters of a sensor model for binaural sensing**

There are two popular ways of determining localisation given binaural cues. One is to establish closed-form expressions function of the interaural baseline and the speed of sound (which disregard the head shadow, and only localise in azimuth); the other is to build the so-called Head-Related Transfer Function (HRTF) by moving a sound source producing white noise (therefore

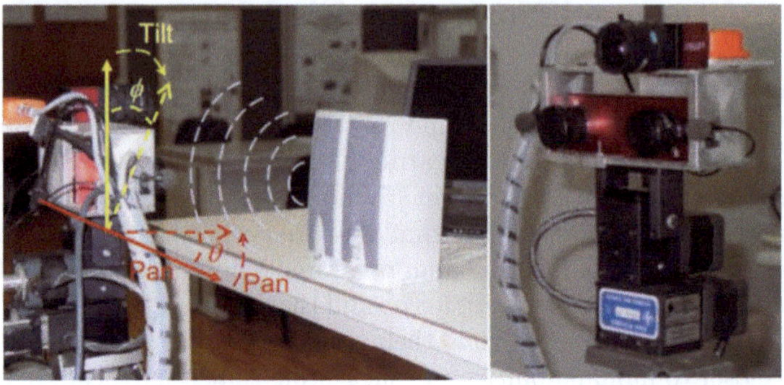

Fig. 6.4. Experimental setup for the binaural sensor model MLE-based learning procedure using the first version of the Integrated Multimodal Perception Experimental Platform (IMPEP).

exploring the full auditory frequency spectrum), in order to sample a spherical grid around the binaural sensing system in terms of ITDs and ILDs.

Ferreira, Pinho, and Dias [4] decided to use a process analogous to the HRTF in order to learn the sensor model likelihoods we have already presented in Example 2.3, by taking several ITD/n-ILD samples for each BVM cell so as to build a statistical description that would allow the MLE of the mean and standard deviation parameters of the $P(\tau \mid S_c)$ and $P(\Delta L(f_c^k) \mid S_c)$, modelled as normal distributions. In the case of $[S_c = 1]$ only samples from cell c occupied by the sound source would be used, while for $[S_c = 0]$ samples from all cells other than c would be used. See Fig. 6.4 for a view of the setup corresponding to the first experiments using this procedure.

Two interesting and illustrative examples which, while not originally used with robots, are easily adaptable to robotic applications would be the work by Hy and Bessière [7], which demonstrates the use of EM and the Baum-Welch algorithm to learn a behaviour model for an AI avatar by human demonstration in an *Unreal Tournament* virtual reality console gaming setting, and the work by Fox [5], which showcases learning using nonparametric hierarchical Bayes models.

6.9 Final Remarks and Further Reading

Learning is a hot topic in robotics research, and as such should be considered by the reader as an important field to investigate further.

Many excellent references exist on probabilistic learning (in particular in the context of graphical models) – see, for example, the textbook by Darwiche [3] or Koller and Friedman [6], the latter being particularly comprehensive and easy to follow, or the seminal publication by Buntine [12]. A good starting point for EM methods and the Baum-Welch algorithm should be to read the original publications [13; 14], and to explore from then on further. The major reference for reinforcement learning would be the survey by Kaelbling, Littman, and Moore [11]. Finally, good references for structure learning methods and nonparametric models would be, as already mentioned, [6; 8; 1].

References

1. Tenenbaum, J.B., Kemp, C., Griffiths, T.L., Goodman, N.D.: How to grow a mind: Statistics, structure, and abstraction. Science 331(6022), 1279–1285 (2011) 165, 167
2. Colas, F., Diard, J., Bessiére, P.: Common Bayesian Models For Common Cognitive Issues. Acta Biotheoretica 58(2-3), 191–216 (2010) 158, 159, 165
3. Darwiche, A.: Modeling and reasoning with Bayesian networks. Cambridge University Press, Cambridge (2009) 154, 156, 163, 166
4. Ferreira, J.F., Pinho, C., Dias, J.: Implementation and Calibration of a Bayesian Binaural System for 3D Localisation. In: 2008 IEEE International Conference on Robotics and Biomimetics (ROBIO 2008), Bangkok, Thailand (2009) 166
5. Fox, E.: Bayesian Nonparametric Learning of Complex Dynamical Phenomena. Ph.D. thesis, MIT, Cambridge, MA (2009) 166
6. Koller, D., Friedman, N.: Probabilistic graphical models: principles and techniques. MIT Press (2009) 152, 154, 159, 165, 166, 167
7. Hy, R.L., Bessiére, P.: Probabilistic Reasoning and Decision Making in Sensory-Motor Systems. In: Bessiére, P., Laugier, C., Siegwart, R. (eds.) Playing to Train Your Video Game Avatar. STAR, vol. 46, pp. 263–278. Springer, Heidelberg (2008) 166
8. Kemp, C., Tenenbaum, J.B.: The discovery of structural form. Proceedings of the National Academy of Sciences 105(31), 10687–10692, 1091–6490 (2008), doi:10.1073/pnas.0802631105 ISSN 0027-8424, PMID: 18669663 165, 167
9. Bergemann, D., Välimäki, J.: Bandit problems. Technical report, Cowles Foundation for Research in Economics, Yale University (2006) 164
10. Florez-Larrahondo, G.: Incremental learning of discrete hidden markov models. Ph.D. thesis, Mississippi State University, Mississippi State, MS, USA. AAI3193417 (2005) 163
11. Kaelbling, L.P., Littman, M.L., Moore, A.W.: Reinforcement Learning: A Survey. Journal of Artificial Intelligence Research 4, 237–285 (1996) 164, 167
12. Buntine, W.L.: Operations for Learning with Graphical Models. Journal of Artificial Intelligence Research (AI Access Foundation) 2, 159–225 (1994) ISSN 11076-9757 167
13. Dempster, A.P., Laird, N.M., Rubin, D.B.: Maximum Likelihood from Incomplete Data via the EM Algorithm. Journal of the Royal Statistical Society Series B (Methodological) 39(1), 1–38 (1977) 162, 167
14. Baum, L.E., Petrie, T., Soules, G., Weiss, N.: A maximization technique occurring in the statistical analysis of probabilistic functions of Markov chains. The Annals of Mathematical Statistics 41(1), 164–171 (1970) 163, 167

Part II

Probabilistic Approaches for Robotic Perception in Practice

Part II

7

Case-Study: Bayesian 3D Independent Motion Segmentation with IMU-Aided RBG-D Sensor

7.1 Introduction

7.1.1 General Goals and Motivations

In this chapter, we will present a case-study consisting of a two-tiered hierarchical Bayesian model to estimate the location of objects moving independently from the observer, reported in the publication by Lobo, Ferreira, Trindade, and Dias [1].

Biological vision systems are very successful in motion segmentation, since they efficiently resort to flow analysis and accumulated prior knowledge of the 3D structure of the scene. Artificial perception systems may also build 3D structure maps and use optical flow to provide cues for ego- and independent motion segmentation. Using inertial and magnetic sensors and a image and depth sensor (RGB-D) the authors proposed a method to obtain registered 3D maps, which are subsequently used in a probabilistic model (the bottom tier of the hierarchy) that performs background subtraction across several frames to provide a prior on moving objects.

The egomotion of the RGB-D sensor is estimated starting with the angular pose obtained from the filtered accelerometers and magnetic data. Its translation is derived from matched points across the images and corresponding 3D points in the rotation compensated depth maps. A gyro-aided Lucas Kanade tracker is used to obtain matched points across the images. The tracked points are also used to refine the initial sensor based rotation estimation. Having determined the camera egomotion, the estimated optical flow assuming a static scene can be compared with the observed optical flow via a probabilistic model (the top tier of the hierarchy), using the results of the background subtraction process as a prior, in order to identify volumes with independent motion in the corresponding 3D point cloud. To deal with the computational load CUDA-based solutions on GPUs were used.

J.F. Ferreira and J. Dias: *Probabilistic Approaches for Robotic Perception*, 171
Springer Tracts in Advanced Robotics 91,
DOI: 10.1007/978-3-319-02006-8_7,

7.1.2 Background

Motion cues play an essential part in perception – they are ubiquitous in the process of making sense of the surrounding world, both for humans and for robots. However, motion perception has been long considered a difficult problem to tackle in artificial perception; although there has been a substantial amount of work in attempting to devise a solution by solely using vision, the challenges faced by the need to distinguish between optical flow caused by self-motion of the observer (i.e. egomotion) and by objects or agents moving independently from the observer are not at all trivial.

In biological vision systems both static and dynamic inertial cues provided by the vestibular system also play an important role in perception. In particular, they are deeply involved in the process of motion sensing, and are fused with vision in the early processing stages of image processing (e.g, the gravity vertical cue). As a result, artificial perception systems for robotic applications have since recently been taking advantage from low-cost inertial sensors for complementing vision systems [5].

On the other hand, an interesting hypothesis has been raised by studies in neuroscience such as presented by Bullier [9], which states that there are fast routes in the brain that are used to rapidly paint the rough overall 3D view of an observed scene, which is then fed back to lower levels of 2D perceptual processing as a prior. In fact, it is also posited by several authors that an accumulated prior knowledge of the 3D structure of the scene is retroinjected into the primary brain sites for flow analysis, thus modulating motion segmentation processing.

Besides the work described in [5] and references therein, recent work had been done in re-examiing the Lucas-Kanade method for real-time independent motion detection [2].

7.2 IMU-Aided RGB-D Sensor for Estimating Egomotion and Registering 3D Maps

7.2.1 Estimating and Compensating for Egomotion

A moving RGB-D observer of a background static scene with some moving objects computes at each instant a dense depth map corresponding to the captured image. The maps will change in time due to both the moving objects and the observer egomotion. A first step to process the incoming data is to register the maps to a common fixed frame of reference $\{\mathcal{W}\}$, as shown on Figure 7.1.

A set of 3D points ${}^{\mathcal{C}}\mathbb{P}|_i$ is therefore obtained at each frame, given in the camera frame of reference $\{\mathcal{C}\}|_i$. Each 3D point has an RGB value and a corresponding intensity gray level c given by the pixel in the reference camera. Each point in the set retains both 3D position and gray level

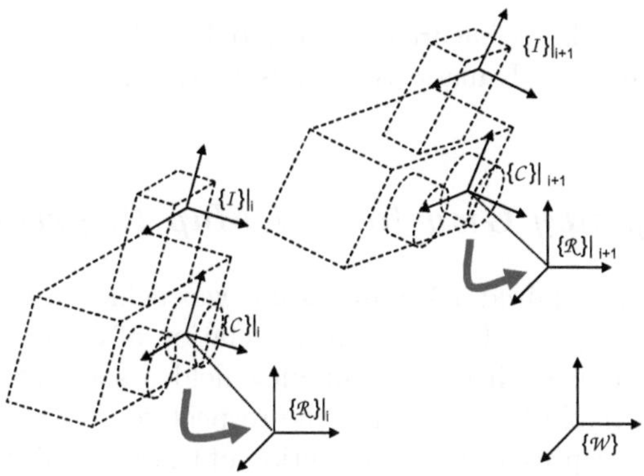

Fig. 7.1. Moving observer and world fixed frames of reference

$$\boldsymbol{P}(x, y, z, c) \in {}^{\mathcal{C}}\mathbb{P}|_i \ . \tag{7.1}$$

The inertial vertical reference alone could be used to rotate depth maps to a levelled frame of reference. However there remains a rotation about a vertical axis for which gravity provides no cues. The earth's magnetic field can be used to provide the missing bearing [10], however the magnetic sensing is sensitive to the nearby ferrous metals and electric currents. In fact, there is some overlap and complementarity between the two sensors, with different noise characteristics that can be exploited to provide a useful rotation update [8; 7].

The inertial and magnetic sensors, rigidly fixed to the depth camera rig, provide a stable camera rotation update ${}^{\mathcal{R}}\boldsymbol{R}_{\mathcal{C}}$ relative to the local gravity vertical and magnetic north camera frame of reference $\{\mathcal{R}\}|_i$. Calibration of the rigid body rotation between $\{\mathcal{I}\}|_i$ and $\{\mathcal{C}\}|_i$ can be performed by having both sensors observing gravity, such as vertical vanishing points and sensed acceleration, as described in [6]. The rotated camera frame of reference $\{\mathcal{R}\}|_i$ is time-dependent only due to the camera system translation, since rotation has been compensated for.

The translation component can be obtained using a single fixed target tracked in the scene, or a set of tracked features to improve robustness. The image features must have the corresponding 3D point $\boldsymbol{P}_t$ in each depth map, so that translation can be estimated from

$$\Delta\mathrm{t} = \boldsymbol{P}_t|_{i+1} - \boldsymbol{P}_t|_i \tag{7.2}$$

with $\boldsymbol{P}_t|_{i+1} \in {}^{\mathcal{R}}\mathbb{P}|_{i+1}$ and $\boldsymbol{P}_t|_i \in {}^{\mathcal{R}}\mathbb{P}|_i$.

A set of sparse tracked natural 3D features can be used to improve robustness, but some assumptions have to be made in order to reject outliers that occur from tracking features of the moving objects. For this work we used a gyro-aided Luca Kanade tracker is used, running on a GPU using CUDA based code [3; 4].

7.2.2 Occupancy Grid for 3D Map Registration

Registration of the acquired 3D point clouds was achieved by using an occupancy grid $\mathcal{Y}$ – a regular 3D Cartesian tesselation of cells (i.e. voxels), each indexed by C, coupled with an occupancy field associating each cell to a binary random variable O_C signalling the respective occupancy state.

Let $Z \equiv \cap_{i=1}^{N} Z_i$ represent the conjunction of the set of discretised readings corresponding to N points (x_i, y_i, z_i) composing the point cloud obtained by the range sensor, assumed to be *conditionally independent measurements*. The occupancy grid is to be updated by inferring $P(O_C \mid Z, M_C, Z_C)$ for each C, through the application of Bayes rule and marginalisation to the standard decomposition equation

$$P(O_C, D, Z, M_C, Z_C) = \\ P(D)P(O_C)P(Z_C \mid O_C, D) \prod_{i=1}^{N} P(M_C^i)P(Z_i \mid M_C^i, O_C, D), \tag{7.3}$$

where $M_C \equiv \cap_{i=1}^{N} M_C^i$ is the conjunction of N random variables M_C^i that signal if the corresponding Z_i falls within the limits of cell C, Z_C signals if there are *any* points within set Z falling within the limits of cell C, and finally D represents a binary random variable signalling either "detection" or "misdetection". The distributions involved in the decomposition are defined in the following lines.

The prior distribution $P([D = 0]) = P_{miss}$, $P([D = 1]) = 1 - P_{miss}$ introduces a meaningful error model that avoids deadlocks caused by 0 or 1 probabilities of occupancy, with P_{miss} being attributed an empirically chosen value; it also establishes the amount of inertia of the model with respect to changing the occupancy state of a cell after consecutive updates of the grid. The distribution $P(O_C)$ represents the prior on occupancy, taken from the posterior estimated in the previous time instant. Each distribution $P(M_C^i)$ represents a uniform (uninformative) prior.

The likelihood $P(Z_i \mid M_C^i, O_C, D)$ represents the direct sensor model of the generative formulation of the occupancy grid given by a delta Dirac distribution displaced to $Z_i = C$ if $M_C^i = 1$ and $D = 1$, or a uniform distribution $\mathcal{U}(Z_i)$ otherwise. Finally, the likelihood $P(Z_C \mid O_C, D)$ represents the probability of $O_C = 0$ implying that no measurement is falling within the limits of cell C; it is given by $P(Z_C|[O_C = 0], [D = 1]) = Z_C$, or a uniform distribution otherwise.

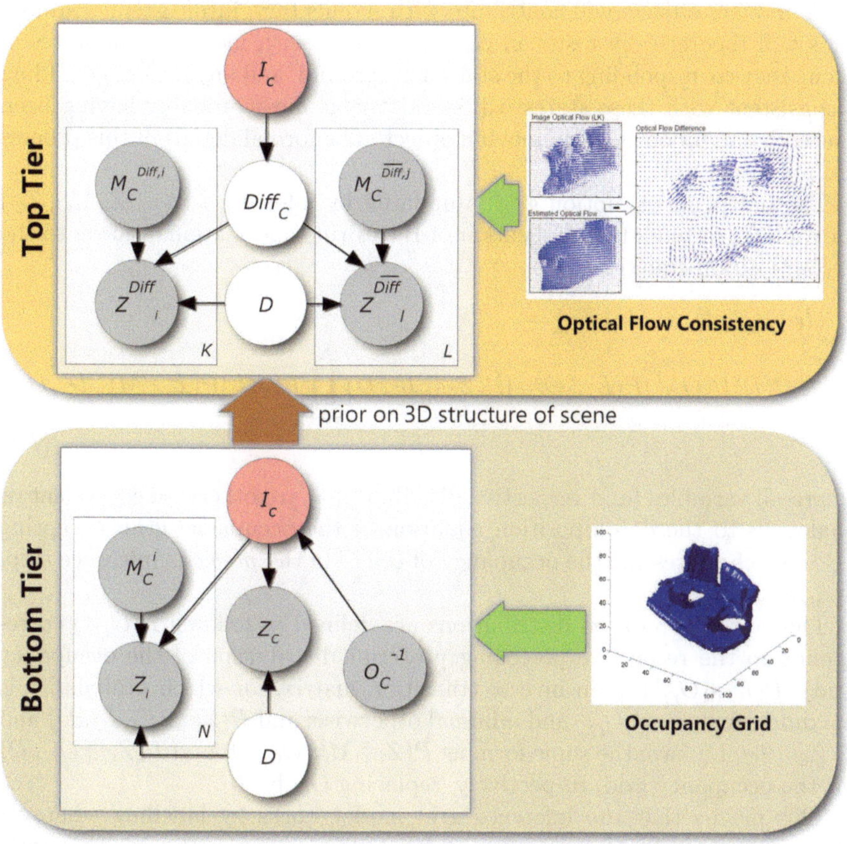

Fig. 7.2. Full hierarchical framework for independent motion segmentation. Bayesian networks using plate notation (Chapter 3) corresponding to each of the hierarchy tiers are presented, with searched variables in red, hidden/unwanted variables to marginalise with no fill and measured variables in grey.

7.3 Two-Tiered Bayesian Hierarchical Model for Independent Motion Segmentation

7.3.1 Bottom Tier – Bayesian Model for Background Subtraction

Background subtraction is performed by updating an inference grid similar to the occupancy grid described in section 7.2.2, but, instead of occupancy, relating to the presence/absence of independent moving objects in cell C, represented by the binary random variable I_C. The rationale of background subtraction in this context is as follows: static objects will contribute with a steady influx of consistent readings registered in the occupancy grid, while moving objects will contribute with momentary, inconsistent readings. This will theoretically result in cells associated with more certain states of occupancy corresponding to the static background, and any incoming reading inconsistent with these states will stand out as most probably having been caused by an independently moving object. The formal details of this process are presented next.

The independent motion grid is updated by inferring $P(I_C \mid Z, M_C, Z_C)$ for each C, through the application of Bayes rule and marginalisation to the decomposition equation

$$P(I_C, O_C^{-1}, D, Z, M_C, Z_C) = \\ P(D)P(O_C^{-1})P(I_C \mid O_C^{-1})P(Z_C \mid I_C, D) \prod_{i=1}^{N} P(M_C^i)P(Z_i \mid M_C^i, I_C, D), \tag{7.4}$$

where all variables (and respective distributions) are otherwise equivalent or analogous to the decomposition equation of the occupancy grid, excepting O_C^{-1}, which represents the occupancy of cell C in the *previous* inference step, and I_C.

The newly introduced distributions are defined as follows: $P(O_C^{-1})$ corresponds to the respective preceding posterior distribution of the occupancy grid); $P(I_C \mid O_C^{-1})$ is an inverse transition matrix, for which probability is maximal when $I_C \neq O_C^{-1}$ and minimal otherwise; and $P(Z_i \mid M_C^i, I_C, D)$ and $P(Z_C \mid I_C, D)$ have the same form as $P(Z_i \mid M_C^i, O_C, D)$ and $P(Z_C \mid O_C, D)$ for the occupancy grid, respectively, replacing O_C by I_C.

This means that the inference grid model works by labelling whatever object perceived by the range sensor that does not comply with the static background that has previously been mapped into the occupancy grid (i.e. $I_C \neq O_C^{-1}$) *as an independently moving object.*

7.3.2 Top Tier – Bayesian Model for Optical Flow Consistency-Based Segmentation

Optical flow is the apparent motion of brightness patterns in the image. Generally, optical flow corresponds to the projected motion field, but not always. Shading, changing lighting and some texture patterns might induce an optical field different from the motion field. However since what can be observed is the optical field, the assumption is made that optical flow field provides a good estimate for the true projected motion field.

Optical flow computation can be performed in a *dense* way, by estimating motion vectors for every image pixel, or it can be *feature-based*, estimating motion parameters only for matched features.

Representing the 2D velocity of an image pixel $\boldsymbol{u} = (u, v)^\mathsf{T}$ as $\frac{d\boldsymbol{u}}{dt}$, the brightness constancy constraint says that the projection of a world point has a constant intensity over a short interval of time, i.e., assuming that the pixel intensity or brightness is constant during dt, we have

$$I(u + \frac{du}{dt}dt, v + \frac{dv}{dt}dt)|_{t+dt} = I(u, v)|_t \tag{7.5}$$

If the brightness changes smoothly with u, v and t, we can expand the left-hand-side by a Taylor series and reject the higher order terms to obtain

$$\nabla \boldsymbol{I} \cdot \frac{d\boldsymbol{u}}{dt} + \frac{\partial I}{\partial t}dt = 0 \tag{7.6}$$

where ∇I is the image gradient at pixel $\boldsymbol{u}$. These spatial and time derivatives can be estimated using a convolution kernel on the image frames.

But for each pixel we only have one constraint equation, and two unknowns. Only the *normal flow* can be determined, i.e., the flow along the direction of image gradient. The flow on the tangent direction of an isointensity contour cannot be estimated. This is the so called *aperture problem.* Therefore, to determine optical flow uniquely additional constraints are needed. The problem is that a single pixel cannot be tracked, unless it has a distinctive brightness with respect to all of its neighbours. If a local window of pixels is used, a local constraint can be added, i.e., single pixels will not be tracked, but windows of pixels instead.

Barron, Fleet, and Beauchemin [11] present a quantitative evaluation of optical flow techniques, including the Lucas-Kanade method, that uses local consistency to overcome the aperture problem [12]. The assumption is made that a constant model can be used to describe the optical flow in a small window.

When the camera is moving and observing a static scene with some moving objects, some optical flow will be consistent with the camera egomotion observing the static scene, other might be moving objects. Since the stereo provides a dense depth map, and we reconstruct camera motion, we can compute the expected projected optical flow in the image from the 3D data.

In the perspective camera model, the relationship between a 3D world point $\boldsymbol{x} = (X, Y, Z)^\mathsf{T}$ and its projection $\boldsymbol{u} = (u, v)^\mathsf{T}$ in the 2D image plane is given by

$$u = \frac{\boldsymbol{P}_1 (x, y, z, 1)^\mathsf{T}}{\boldsymbol{P}_3 (x, y, z, 1)^\mathsf{T}} \qquad v = \frac{\boldsymbol{P}_2 (x, y, z, 1)^\mathsf{T}}{\boldsymbol{P}_3 (x, y, z, 1)^\mathsf{T}} \tag{7.7}$$

where matrix $\boldsymbol{P}_j$ is the jth row of the camera projection matrix $\boldsymbol{P}$.

When the camera moves, the relative motion of the 3D point $\frac{d\boldsymbol{x}}{dt}$ will induce a projected optical flow given by

$$\frac{d\boldsymbol{u}_i}{dt} = \frac{\delta \boldsymbol{u}_i}{\delta \boldsymbol{x}} \frac{d\boldsymbol{x}}{dt} \tag{7.8}$$

where $\frac{\delta \boldsymbol{u}_i}{\delta \boldsymbol{x}}$ is the 2×3 Jacobian matrix that represents the differential relationship between $\boldsymbol{x}$ and $\boldsymbol{u}_i$, which can be obtained by differentiating (7.7).

Image areas where the computed flow is inconsistent with the expected one indicate moving objects, and the corresponding voxels can be segmented.

The difference image between the estimated and the measured optical flow is then thresholded and binarised. Consequently, two mutually exclusive sets of random variables of the same form as Z can be defined, Z^{Diff} and $Z^{\overline{Diff}}$, by classifying points from the cloud yielded by the range sensor as either corresponding to a *non-consistent pixel* or to a *consistent pixel* with corresponding variables analogous to M_C, M_C^{Diff} and $M^{\overline{Diff}_C}$, respectively.

Using these random variables, the top-level inference grid is updated by inferring $P(I_C \mid Z^{Diff}, Z^{\overline{Diff}}, M_C^{Diff}, M_C^{\overline{Diff}})$ for each C, through the application of Bayes rule and marginalisation to the decomposition equation

$$\begin{aligned} P(I_C, Z^{Diff}, Z^{\overline{Diff}}, M_C^{Diff}, M_C^{\overline{Diff}}, D, Diff_C) = \\ P(D)P(I_C)P(Diff_C \mid I_C) \\ \prod_{i=1}^{K} P(M_C^{Diff,i})P(Z_i^{Diff} \mid Diff_C, M_C^{Diff,i}, D) \\ \prod_{j=1}^{L} P(M_C^{\overline{Diff},j})P(Z_j^{\overline{Diff}} \mid Diff_C, M_C^{\overline{Diff},j}, D), \end{aligned} \tag{7.9}$$

where the remaining random variables have the same meaning as before, with the exception of $Diff_C$, a hidden binary variable which signals if a cell C is labelled as being occupied by an independently moving object, *if considering consistency-based segmentation.*

Since it is expected that consistency-based segmentation and background subtraction segmentation yield the same results, the distribution $P(Diff_C \mid I_C)$ is simply a transition matrix for which probability is maximal when $Diff_C = I_C$ and minimal otherwise. The distribution $P(I_C)$ provides *the link between the two tiers of the hierarchy* by applying the probabilistic subroutine concept presented in Chapter 4, and is given by the result of inference on the lower level, $P(I_C \mid Z, M_C, Z_C)$. It models the *accumulated prior knowledge of the 3D structure of the scene*, thus representing an analogous process to what is believed to happen in the human brain, as described in the introductory section.

Finally, $P(D)$ and $P(M_C^{Diff,i})$, $P(M_C^{\overline{Diff},j})$ and $P(Z_i^{Diff} \mid Diff_C, M_C^{Diff,i}, D)$ follow analogous definitions to the corresponding distributions in previous models, while $P(Z_j^{\overline{Diff}} \mid Diff_C, M_C^{\overline{Diff},j}, D)$ is given by a delta Dirac distribution displaced to $Z_j^{\overline{Diff}} = C$ for $Diff_C = 0$, $M_C^{\overline{Diff},j} = 1$ and $D = 1$, or a uniform distribution $\mathcal{U}(Z_i)$ otherwise.

The full hierarchical framework is presented on Fig. 7.2. The posterior of the top tier of the hierarchy only needs to be inferred up to a proportion of the product of the nonuniform priors and likelihoods, to then apply a maximum a posteriori (MAP) decision rule in order to estimate the segmented independent motion. Conversely, the posterior distributions of the occupancy grid and the bottom tier of the hierarchy should be exactly inferred; however, the respective models have been designed so that inference can be easily and efficiently performed using closed-form solutions. The derivations of the expressions involved in these processes are presented next.

7.4 Closed-Form Derivations of Inference and MAP Estimation Expressions

In the following derivations, consider the shorthand notations $d \equiv [D = 1]$ and $\neg d \equiv [D = 0]$, $o_C \equiv [O_C = 1]$ and $\neg o_C \equiv [O_C = 0]$, and $i_C \equiv [I_C = 1]$ and $\neg i_C \equiv [I_C = 0]$.

Starting with the marginalisation of D in equation (7.3), we have

$$\begin{aligned} &P(O_C, Z_1, \ldots, Z_N, Z_C, M_1^C, \ldots, M_N^C) = \\ &\quad \sum_D P(O_C, Z_1, \ldots, Z_N, Z_C, M_1^C, \ldots, M_N^C) = \\ &\quad \sum_D P(D)P(O_C)P(Z_C \mid O_C, D) \prod_{i=1}^{N} P(M_C^i)P(Z_i \mid M_C^i, O_C, D). \end{aligned} \tag{7.10}$$

Given that $P(M_C^i)$ are uniform priors, and that both these distributions and $P(O_C)$ are independent of D, they can be factored out of the marginalisation summation expression. Assuming N_C as representing the total number of measurements of set Z *effectively* influencing cell C, one can substitute the following expression

$$Term_{\neg d} = .5 \times P_{miss} \mathcal{U}(Z_i)^N = 5. \times P_{miss} \mathcal{U}(Z_i)^{N-N_C} \overbrace{\mathcal{U}(Z_i)^{N_C}}^{\Phi},$$

in equation 7.10, thereby obtaining

$$\begin{aligned} &P(O_C, Z_1, \ldots, Z_N, Z_C, M_1^C, \ldots, M_N^C) = \\ &\quad P(O_C)P(M_C^i)^N \times \\ &\quad \left((1 - P_{miss})P(Z_C \mid O_C, d) \prod_{i=1}^{N} P(Z_i \mid M_C^i, O_C, d) + Term_{\neg d} \right) = \\ &\quad P(O_C)P(M_C^i)^N \times \\ &\quad \left((1 - P_{miss})P(Z_C \mid O_C, d) \prod_{i=1}^{N} P(Z_i \mid M_C^i, O_C, d) + Term_{\neg d} \right). \end{aligned} \tag{7.11}$$

Let us now assume

$$A = P(o_C)\left(.5(1-P_{miss}) \times 1^{N^C} \times \mathcal{U}(Z_i)^{N-N_C} + .5P_{miss}\mathcal{U}(Z_i)^{N-N_C}, \overbrace{\mathcal{U}(Z_i)^{N_C}}^{\Phi}\right),$$

and

$$B = P(\neg o_C)\left[Z_C(1-P_{miss})\mathcal{U}(Z_i)^{N-N_C}\overbrace{\mathcal{U}(Z_i)^{N_C}}^{\Phi} + .5P_{miss}\mathcal{U}(Z_i)^{N-N_C}\overbrace{\mathcal{U}(Z_i)^{N_C}}^{\Phi}\right].$$

The closed-form solution for exact inference of $P(o_C \mid \ldots)$ after marginalisation of D in equation (7.3) then becomes

$$P(o_C \mid Z_1,\ldots,Z_N,Z_C,M_1^C,\ldots,M_N^C) = \frac{A}{A+B} =$$
$$\frac{\overbrace{P(o_C)\left(.5(1-P_{miss}) + .5P_{miss}\Phi\right)}^{A'}}{\underbrace{P(o_C)\left(.5(1-P_{miss}) + .5P_{miss}\Phi\right)}_{A'} + P(\neg o_C)\left[(Z_C(1-P_{miss}) + .5P_{miss})\Phi\right]} =$$
$$\frac{A'}{A' + [1-P(o_C)]\left[(Z_C(1-P_{miss}) + .5P_{miss})\Phi\right]} \tag{7.12}$$

The closed-form solution for exact inference using equation 7.4 is identical; one only needs to replace $P(o_C)$ by $P(\neg o_C)$.

Finally, applying an analogous rationale, the result of marginalisation of D in equation (7.9) and of computing the MAP estimate for $[I_C = 1]$ is given by

$$\begin{aligned} &P(i_C \mid Z^{Diff}, Z^{\overline{Diff}}, M_C^{Diff}, M_C^{\overline{Diff}}, Diff_C) = \\ &\sum_D P(i_C \mid Z^{Diff}, Z^{\overline{Diff}}, M_C^{Diff}, M_C^{\overline{Diff}}, D, Diff_C) \propto \\ &P(i_C)\left[(1-P_{miss}) \times \mathcal{U}(Z_i)^{L+L^C-K^C} + P_{miss}\right]. \end{aligned} \tag{7.13}$$

All of these closed-form expressions were applied simultaneously for all cells in a SIMD (single-instruction, multiple-data) parallel-programming implementation in the spirit of the algorithms introduced in Appendix A, in a similar fashion as is presented in the following chapter.

7.5 Experimental Results

Using a MS Kinect as the RGB-D sensor, and attaching a Xsens MTix IMU sensor, that has both inertial and magnetic sensors, we were able to acquire

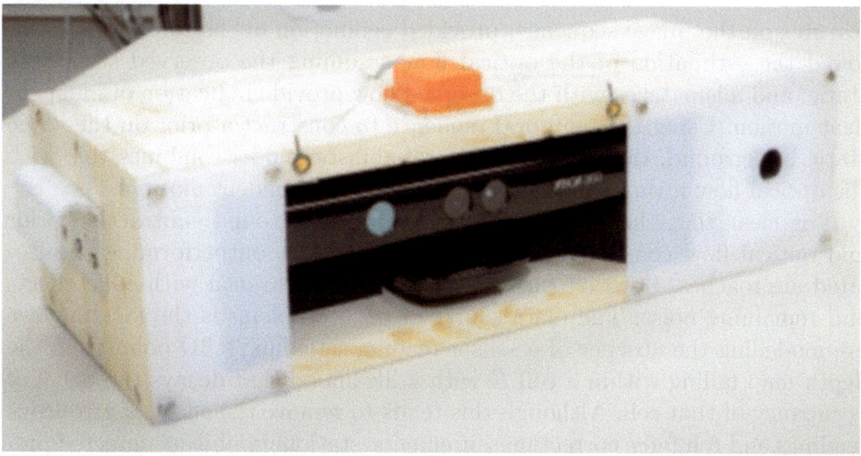

Fig. 7.3. Experimental setup with RGB-D (MS Kinect) and IMU (Xsens MTix) sensors

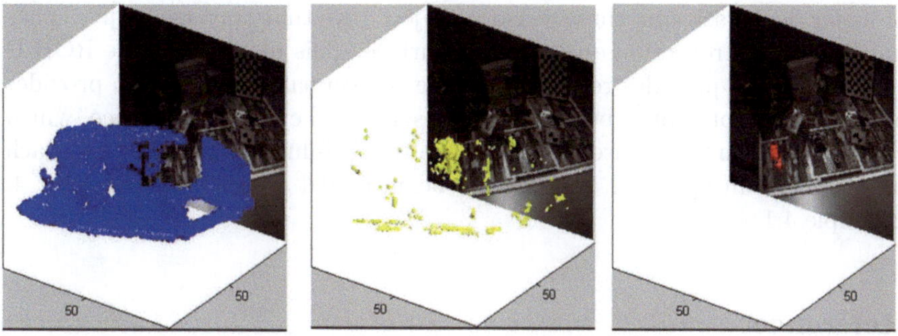

Fig. 7.4. Results showing background subtraction prior (blue on the left), optical flow consistency bottom tier (yellow on the centre) and the final top tier result (red on the right)

datasets with images, depth maps and rotation update. Fig. 7.3 shows the experimental setup that was used, where optotracker markers where added to provide egomotion ground truth, to be used later for benchmarking and refining the implemented method.

Figure 7.4 shows preliminary results of this ongoing work.

7.6 Conclusions and Future Work

In this chapter we presented a case-study consisting of a two-tiered hierarchical Bayesian model to estimate the location of objects moving independently from the observer. Having a RGB-D sensor with and attached IMU we were able to have a rotation update form the IMU, that combined with tracked features on the image sequence, provided egomotion of the sensors. This allowed the estimation of the optical flow assuming the observed scene was static, and mismatches with the observed flow provide indication of independent motion. Using the temporal sequence to construct a prior on the scene static background, the implemented probabilistic model combines this with the optical flow mismatch to find voxels with independent motion.

It is clear that the probabilistic fusion of background subtraction prior and optical flow consistency works to some extent, outperforming the isolated approaches. However further work is needed to deal with edge effects and remaining noise. The main source of both problems is the fact that we are modelling the absence of a sensor reading signalling a 3D point from the depth map falling within a cell C with a likelihood that decays the belief of occupancy of that cell. Although this tends to remove the effect of erroneous readings and reinforce correct measurements, stationary objects detected previously which subsequently fall outside the field of view (i.e., due to sensor egomotion) will eventually be "forgot" by the model.

However, the depth sensor used in this work functions, in fact, as an array of linear depth sensors; these sensors project rays that traverse empty space until there is a reflection on an object surface. This means that the RGB-D sensor not only provides readings relating to occupancy, but it also provides evidence of empty space between the sensor and the detected surface, which could be used in the future to replace the "forgetfulness" likelihood approach by means of an approach such as the beam model presented in Chapter 4, Example 4.1.

References

1. Lobo, J., Ferreira, J.F., Trindade, P., Dias, J.: Bayesian 3D Independent Motion Segmentation with IMU-Aided RBG-D Sensor. In: Proceedings of the 2012 IEEE International Conference on Multisensor Fusion and Information Integration (MFI 2012), Hambur (2012) 171
2. Ciliberto, C., Pattacini, U., Natale, L., Nori, F., Metta, G.: Reexamining Lucas-Kanade method for real-time independent motion detection: Application to the iCub humanoid robot. In: 2011 IEEE/RSJ International Conference on Intelligent Robots and Systems (IROS), pp. 4154–4160 (2011), doi:10.1109/IROS.2011.6094985 172
3. Hwangbo, M., Kim, J.S., Kanade, T.: Inertial-aided klt feature tracking for a moving camera. In: IEEE/RSJ International Conference on Intelligent Robots and Systems, IROS 2009, pp. 1909–1916 (2009), doi:10.1109/IROS.2009.5354093 174
4. Kim, J.S., Hwangbo, M., Kanade, T.: Realtime affine-photometric klt feature tracker on gpu in cuda framework. In: 2009 IEEE 12th International Conference onComputer Vision Workshops (ICCV Workshops), pp. 886–893 (2009), doi:10.1109/ICCVW.2009.5457608 174

5. Corke, P., Lobo, J., Dias, J.: An introduction to inertial and visual sensing. The International Journal of Robotics Research (IJRR) Special Issue from the 2nd Workshop on Integration of Vision and Inertial Sensors 26(6), 519–535 (2007), doi:10.1177/0278364907079279 172
6. Lobo, J., Dias, J.: Relative pose calibration between visual and inertial sensors. The International Journal of Robotics Research (IJRR) Special Issue from the 2nd Workshop on Integration of Vision and Inertial Sensors 26, 561–577 (2007) 173
7. Roetenberg, D., Luinge, H., Baten, C., Veltink, P.: Compensation of magnetic disturbances improves inertial and magnetic sensing of human body segment orientation. IEEE Transactions on Neural Systems and Rehabilitation Engineering 13(3), 395–405 (2005), doi:10.1109/TNSRE.2005.847353, see also IEEE Trans. on Rehabilitation Engineering 173
8. Roetenberg, D., Luinge, H., Veltink, P.: Inertial and magnetic sensing of human movement near ferromagnetic materials. In: The Second IEEE and ACM International Symposium on Mixed and Augmented Reality, pp. 268–269 (2003) 173
9. Bullier, J.: Integrated model of visual processing. Brain Research Reviews 36(2-3), 96–107 (2001) ISSN 0165-0173, doi:10.1016/S0165-0173(01)00085-6 172
10. Caruso, M.J., Bratland, T., Smith, C.H., Schneider, R.: A New Perspective on Magnetic Field Sensing. Technical report, Honeywell, Inc. (1998) 173
11. Barron, J., Fleet, D., Beauchemin, S.: Performance of Optical Flow Techniques. International Journal of Computer Vision 12(1), 43–77 (1994), doi:10.1007/BF01420984 177
12. Lucas, B.D., Kanade, T.: An Iterative Image Registration Technique with an Application to Stereo Vision. In: Proceedings of Imaging Understanding Workshop, pp. 674–679 (1981) 177

8

Case-Study: Bayesian Hierarchy for Active Perception

8.1 Introduction

8.1.1 General Goals and Motivations

Consider the following scenario (Fig. 8.1) – a moving observer is presented with a non-static 3D scene containing several moving entities, probably generating some kind of sound: how does this observer perceive the 3D structure, motion trajectory and velocity of all entities in the scene, while taking into account the ambiguities and conflicts inherent to the perceptual process?

We will present a complex artificial active perception system that follows human-like bottom-up driven behaviours using vision, audition and vestibular sensing, reported on previously published material [1; 2; 4; 5; 8; 13]. This system:

- Deals with perceptual uncertainty and ambiguity, offering some adaptive ingredients that form a reasonable bioinspired basis for a full-fledged robotic perception system.
- Deals with multimodality by tackling sensor fusion geometry in a natural way, consistent with most of the inherent properties of sensation.
- Allows for fast processing of perceptual inputs to build a spatial representation for active perception in a behaviourally relevant fashion, as required in applications in which complex human-robot interaction is required.

8.1.2 Constraining the Problem

The epistemological problem presented in Fig. 8.1 has an incredibly broad reach, and researchers have been incapable of producing a complete and convincing solution for it until the present day. The solution for this problem resides, we believe, in the correct assessment of *the means to make it tractable.*

Nature, through the human brain, provides crucial hints on how this can be done. Biological perception systems do not build representations of their

J.F. Ferreira and J. Dias: *Probabilistic Approaches for Robotic Perception*, Springer Tracts in Advanced Robotics 91,
DOI: 10.1007/978-3-319-02006-8_8,

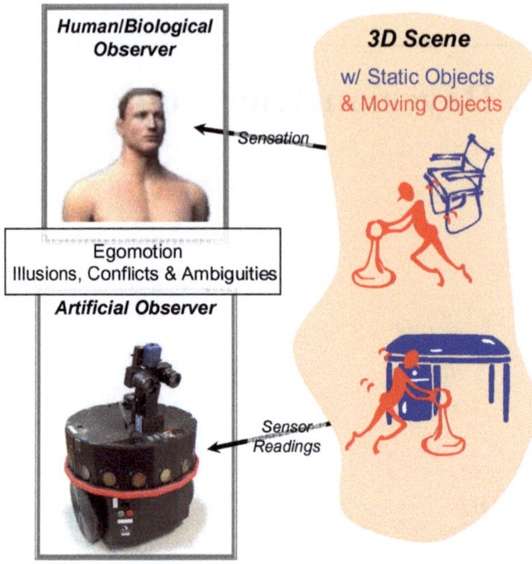

Fig. 8.1. Setting for the perception of 3D structure, ego- and independent motion

surroundings from single instantaneous sensorial snapshots – sensors sweep the perceptual scene, thus changing the focus of attention in an incremental, dynamic fashion. Since humans are prevalently social beings, their attentional system is inherently socially driven; this becomes particularly important when considering human-machine interaction, where robots are expected to engage with humans while displaying attentional behaviours that resemble those of their interlocutors. Even when dealing with unknown environments with no social intent, humans use their own attentional system to its full. For example, a random exploratory strategy alone would not take into account potential primal dangers lurking in our surroundings. In fact, human evolution has genetically imprinted as prior knowledge that certain stimuli are a tell-tale of the proximity of predators, or are caused by competitors of our own species. Consequently, Kopp and Gärdenfors [39] posit that the capacity of attention, and therefore active perception, is a *minimal criterion of intentionality* for robots.

Therefore, it is our belief that the insight provided by the way the human tackles this problem will prove to be invaluable for designing complete, robust perceptual models. We will investigate these matters in the following subsection.

8.1.3 How Does Nature Do It? – Our Black Box

Humans and other animals actively direct their sensors to unknown and also to interesting parts of the perceptual scene, so as to build up a mental map of the surrounding environment. One reason for saccades (i.e. rapid head-eye movements) is to move the senses so that redundant evidence can be accumulated about a scene, lowering the overall uncertainty of individual sensor measurements and using limited-scope sensorial resources more efficiently. In fact, although vision is arguably the main instigator of active perception in mammals, this process is undoubtedly multisensory. As a remarkable example, audition is able to drive gaze shifts towards targets outside the visual field, an ability that has made this sense paramount for the interaction between humans and their surroundings.

The perceptual process is inherently highly intricate and complex. To deal with this challenge, it is believed that, during evolution of the animal brain, this process has been decomposed into simpler subtasks, carried out by modular sites intertwined in a complex fashion, forming a myriad of forward, lateral, and feedback connections. It would be difficult to assume that natural cognitive systems process their complex sensory inputs in a single layer of computation [7; 6]. Therefore perception can be decomposed into subprocesses that communicate intermediate results, which introduces the notion of *modularity*.

Ever since seminal work by Marr [58] up until more recent accounts such as Ballard [44] and many others on computational theories of perception, the link between the functional organisation of perceptual sites in the brain and the underlying computational processes has led to the notion that modularity plays a major role in making these processes tractable. As a matter of fact, although the interconnections between these sites have increasingly been found to be much more intricate than Marr believed, the notion that the brain is organised in a modular fashion is undisputed.

For more than 20 years now, evidence has been accumulating from studies involving healthy human subjects that suggests parallel streams for visual processing for perception versus visual processing for the control of action [38]. In fact, in the human brain, mainly two pathways or streams, anatomically separate albeit interconnected in a complex fashion, have been found to be involved in sensory processing: the *dorsal pathway* and the *ventral pathway*.

Two main theories have arisen over the exact nature of the function of these two pathways, depending on whether emphasis is placed on the input distinctions or on output requirements. Over 20 years ago, Ungerleider and Mishkin [59; 57] described the functions of the two cortical systems based on the former, as a distinction between "object" versus "spatial" vision. Based on the latter, on the other hand, circa 10 years later, Goodale and Milner [53] advanced the argument that the distinction is perhaps more parsimoniously described as one between visual "perception" and the visual control of "action". In this more recent account, the ventral stream of visual projections mediates the perception of objects and their relations, whereas the dorsal stream mediates the visual control of actions directed to these objects [48].

In either case, it is consensual that the dorsal stream, commonly called the "Where" or "How Pathway" depending on the theory, is associated with motion, representation of object locations, and control of the eyes and arms, especially when visual information is used to guide saccades or reaching, and that the ventral stream, commonly called the "What Pathway", is associated with form recognition and object representation. The latter is additionally believed to be associated with storage of long-term memory. It is also consensual nowadays that the widespread interconnections between the two pathways imply that their performances are strongly correlated in the undamaged brain (see, for example, Dyde and Milner [38]), while decorrelation is more evident in clinical models of dorsal stream dysfunction (see Castelo-Branco et al. [21]).

It is believed that there are multimodal perceptual feedback loops stemming from other sensory cortices and processing regions in the brain to the visual pathways. On the other hand, it is known that an additional sensory processing site is heavily permeated by multimodal signals: the *superior colliculus* (SC).

These findings support the construction of a framework that allows fast processing of perceptual inputs to build a perceptual map of space so as to promote immediate action on the environment (as in the dorsal stream and superior colliculus), effectively postponing data association such as object segmentation and recognition (as in the ventral stream) to higher-level stages of processing – this would be analogous to a tennis player being required to hit a ball regardless of perception of its texture properties.

A saccade is a fast movement of an eye, head or other part of an animal's body or device. For example, eye saccades are quick, simultaneous movements of both eyes in the same direction. Saccades serve several purposes, such as a mechanism for fixation or rapid eye movement [23].

Visual saccades, the most thoroughly investigated type of saccade, are measured or investigated in four ways (which can be generalised to multisensory-driven saccades) [20]:

- In a visually guided saccade, an observer performs a *gaze shift* towards a visual onset, or stimulus. This is typically included as a baseline when measuring other types of saccades.
- In an antisaccade, an observer moves eyes away from the visual onset. They are more delayed than visually guided saccades, and observers often make erroneous saccades in the wrong direction. A successful antisaccade requires inhibiting a reflexive saccade to the onset location, and voluntarily moving the eye in the other direction.
- In a memory guided saccade, an observer shifts its gaze towards a remembered point, with no sensory onset involved.
- In smooth pursuit eye movements, an observer tracks a small object moving with a constant slow speed. They emphasise basic eye control, not cognitive processes.

Sensory guided and memory guided saccades involve *gaze computation*, the object of the models presented herewith, followed by *gaze control*, which translates desired fixation points to sequences of commands to the eye and head (i.e. motor commands) and is beyond the scope of this text. Gaze computation is typically broken up into two phases: an *attention model* that identifies relevant features in the scene, selects one of these features and maintains focus on it, and a *gaze policy*, that operates over the feature map to determine the actual fixation point [27; 39].

There are many ways that can be used to classify an attention system according to its various aspects. In a subject's point of view, gaze fixation may be switched to the point being attended to (i.e., overt attention) or, alternatively, attentional processing may also be switched without involving any fixation shift or motor action (i.e., covert attention). We will be focusing our attention on the former.

On the other hand, in order that behaviourally relevant perceptual information is appropriately selected, efficient mechanisms must be in place. Two major attentional mechanisms are known to control this selection process [40]. First, bottom-up attentional selection is a fast, and often compulsory, stimulus-driven mechanism (related to the so-called *exogenous attention*). Research has been proven that attention can be captured under the right stimulus conditions. For example, highly salient feature singletons or abrupt, unexpected onsets attract attention (pop-up effects). On the other hand, top-down attentional selection, is a slower, goal-directed mechanism, where the observer's expectations or intentions influence the allocation of attention (related to the so-called *endogenous attention*). Observers can voluntarily select regions of space or individual objects to attend. The degree to which these two mechanisms influence attentional selection under natural viewing conditions has been for a long time under debate [40].

A great deal of research has been dedicated to developing models of visual attention in the past few years. These computational models are just rough approximations to the human visual attention system and typically operate by identifying, within an incoming visual stream, spatial points of interest. This computational formulation of perceptual attention is very limiting, in terms of the capabilities and complexities of the biological reality [27]. These models serve to reduce the scene to several points of particular interest, and to emulate the scan-path behaviour of human subjects. In this fashion, it is possible to control the combinatorial explosion that results from the consideration of all possible image relationships and provide a naturalistic interface to behaviours such as joint attention [27].

However, even in visual animals *multisensory* stimuli (e.g. visual, auditory or tactile) elicit gaze shifts to aid visual perception of stimuli. Such gaze shifts can either be top-down attention driven (e.g. visual search) or they can be reflex movements triggered by unexpected changes in the surroundings triggered by the collective result of multimodal perception [24].

Several representative models addressing most of these issues will be briefly reviewed in the following lines.

One of the most popular computational models serving as a basis for robotic implementations of visual attention is the model by Itti et al. [47].

This model has roots at least as far back as [50] and its most recent developments are described in [29].

Itti et al.'s model is a feed-forward bottom-up computational model of visual attention, employing, at its most basic level, decompositions into purely preattentive features. Reportedly, this offers advantages in both speed and transparency. As described by Shic and Scassellati [27] in their survey, it is a model that is not only simple but also rigorously and specifically defined, making it a strong contender in terms of implementation, extension, and reproducibility of results. The model extracts the preattentive modalities of colour, intensity, and orientation from an image. These modalities are assembled into a multiscale representation using Gaussian and Laplacian pyramids. Within each modality, centre-surround operators are applied in order to generate multiscale feature maps. An approximation to lateral inhibition is then employed to transform these multiscale feature maps into conspicuity maps, which represent the saliency of each modality. Finally, conspicuity maps are linearly combined to determine the saliency of the scene. Although this model did not originally attend to visual motion, known to be a major modality in visual attention, it has been extended to include it in later work, such as [27].

Parkhurst, Law, and Niebur [40] show that the saliency maps of images, as computed by the Itti model, display higher values in locations fixated upon by human subjects than would have been expected by chance alone. The fact that the saliency maps generated by the same computational attention model can be correlated to approximate probability density maps of humans is shown by Ouerhani et al. [35]. The model by Itti et al. is not uncontroversial, as can be seen in the completely different evaluations by Parkhurst et al. [40], who generally validate the model, and Turano, Geruschat, and Baker [36], who attempt to detract from it by claiming that the predicted gaze locations are no better than random, or Tatler et al. [32], who claim that the model is not scale or rotation invariant, thus questioning the appropriateness of using it as the basis of computational object recognition systems[1]. In any case, Itti and coworkers have shown that interesting objects seem to be visually salient, indicating that selecting interesting objects in the scene is largely constrained by low-level visual properties rather than solely determined by higher cognitive processes [18]. Shic and Scassellati [27] build upon the Itti model to apply a framework, based on dimensionality-reduction over the features of human gaze trajectories, that can simultaneously be used for both optimising a particular computational model of visual attention and for evaluating its performance in terms of similarity to human behaviour.

Alternative computational models of visual attention both with and without motion besides the model presented above exist, such as the work of Tsotsos et al. [51] or Breazeal and Scassellati [45] and many others.

The gaze computation process takes, as an input, the saliency map, and returns, as an output, a point of fixation. One of the simplest gaze policies

[1] For a deeper insight, please refer to [27].

that can by employed is to simply index the location in the saliency map corresponding to the highest peak [27].

On the other hand, regarding the *temporal* dimension of attention, a commonly used complementary model is the Inhibition of Return (IoR) mechanism [50]. The IoR, in simple terms, is the mechanism where the saccade generating system in the brain avoids fixation sites which have just been a focus of attention, therefore preventing deadlocks. Recently, a more complex model has been devised, using Bayesian surprise as a factor related to the attentional changes in the time domain, by Itti and Baldi [16].

8.1.4 How Can Robotic Perception Systems Do It?

Active perception has been an object of study in robotics for decades now, specially active vision, which was first introduced by Bajcsy [56] and later explored by Aloimonos, Weiss, and Bandyopadhyay [55]. Many perceptual tasks tend to be simpler if the observer actively shifts attention by controlling its sensors [55]. Active perception is thus an intelligent data acquisition process driven by the measured, partially interpreted scene parameters and their errors from the scene. The active approach has the important advantage of making most ill-posed perception tasks tractable [55].

One of the most popular computational models serving as a basis for robotic implementations of visual attention is the *saliency model* by Itti, Koch, and Niebur [47], described previously. On the other hand, regarding the *temporal* dimension of attention, a commonly used complementary model is the *Inhibition of Return* (IoR) mechanism [50]. The IoR, as described above, is the mechanism where the saccade generating system in the brain avoids fixation sites which have just been a focus of attention, therefore preventing deadlocks and infinite loops.

As discussed in Chapter 4, hierarchical Bayesian methods provide the adequate framework for implementing modularity in perception. The focus of this Chapter will be on the description of the application of the Bayesian Programming formalism (Chapter 3) to develop a hierarchical modular probabilistic framework that allows the combination of active perception behaviours, namely:

- active exploration based on entropy developed in previously published work, using a Bayesian filter operating upon a log-spherical occupancy grid, which, while not strictly neuromimetic, finds its roots in the role of the dorsal perceptual pathway and superior colliculus of the human brain – refer to Chapters 2 and 3 for more details;
- automatic orientation based on sensory saliency [50], also operating upon the same log-spherical grid.

A real-time implementation of all the processes of the framework has been developed, capitalising on the potential for parallel computing of most of its algorithms.

An overview of the framework and its models will be summarised in this text, and results will be presented. In the process, we will demonstrate the following properties which are intrinsic to the framework: *emergence*, *scalability* and *adaptivity*.

Recent work in active vision by Tsotsos and Shubina [28] and Bohg, Barckholst, Huebner, Ralph, Rasolzadeh, Song, and Kragic [9], the former for target search and the latter for object grasping, contrary to our solution, use an explicit representation for objects to implement active perception. On the other hand, several solutions for target applications similar to ours avoid explicit object representation by resorting to a bottom-up saliency approach such as defined by Itti, Koch, and Niebur [47] – examples of these would be Shibata, Vijayakumar, Conradt, and Schaal [42], Breazeal, Edsinger, Fitzpatrick, and Scassellati [41] and Dankers, Barnes, and Zelinsky [22]. Finally, Dankers, Barnes, and Zelinsky [30] use an approach similar to ours, with an egocentric three-dimensional occupancy grid for integrating range information using active stereo and a Bayesian approach, also detecting 3D mass flow. However, this solution suffers from the downside of using an Euclidean tesselation of space, which complicates sensor models for map updating and fixation computation due to the compulsory use of ray-tracing methods. These works, as most solutions in active perception, use a behavioural approach; an alternative is a probabilistic approach that attempts to reduce uncertainty on a part of the world state, modelled as belief [11]. Our work intends to combine both variants into a coherent, albeit more powerful approach.

Active multisensory perception using spatial maps has, contrastingly, been the object of study since only much recently. Few other explicit models exist, although many artificial perception systems include some kind of simple attention module that drives gaze towards salient auditory features. As an example of a full-fledged multisensory attention model, Koene, Morén, Trifa, and Cheng [24] present a general architecture for the perceptual system of a humanoid robot featuring multisensory (audiovisual) integration, bottom-up salience detection, top-down attentional feature gating and reflexive gaze shifting, which is of particular relevance to our work. The complete system focuses on the multisensory integration and desired gaze shift computation performed in the "Superior Colliculus (SC)" module [24]. This allows the robot to orient its head and eyes so that it can focus its attention on audio and/or visual stimuli. The system includes mechanisms for bottom-up stimulus salience based gaze/attention shifts (where salience is a function of feature contrast) as well as top-down guided search for stimuli that match certain object properties. In order to facilitate interaction with dynamic environments the complete perceptual-motor system functions in real-time [24].

The approach presented in this chapter implements active visuoauditory perception, adding to it vestibular sensing/proprioception so as to allow for sensor fusion given a rotational egomotion. However our solution differs from purely saliency-based approaches in that it also implements an active exploration behaviour based on the entropy of the occupancy grid, so as to promote gaze shifts to regions of high uncertainty.

8.2 From Sensation to Perception

8.2.1 Bayesian Framework for Sensor Fusion

A spatial representation framework for multimodal perception of 3D structure and motion, the Bayesian Volumetric Map (BVM), was presented in Chapters 2 and 3. This framework effectively provides a computational means of storing and updating a perceptual spatial map in a short-term working memory data-structure, representing both 3D structure and motion, without the need for any object segmentation process, finding its roots in the role of the superior colliculus and the dorsal perceptual pathway of the human brain.

Summarising what was presented on Chapter 2, in the BVM framework, cells of a partitioning grid on the BVM log-spherical space $\mathcal{Y}$ associated with the egocentric coordinate system $\{\mathcal{E}\}$ are indexed through $C \in \mathcal{Y}$, representing the subset of positions in $\mathcal{Y}$ corresponding to the "far corners" $(\log_b \rho_{\max}, \theta_{\max}, \phi_{\max})$ of each cell C, O_C is a binary variable representing the state of occupancy of cell C (as in the commonly used occupancy grids – see Elfes [54]), and V_C is a finite vector of random variables that represent the state of all local motion possibilities used by the prediction step of the Bayesian filter associated to the BVM for cell C, assuming a constant velocity hypothesis, as depicted on Fig. 8.2. Sensor measurements (i.e. the result of visual and auditory processing) are denoted by Z – observations $P(Z|O_C\, V_C\, C)$ are given by the Bayesian sensor models of Fig. 8.2, which yield results already integrated within the log-spherical configuration.

The complete set of variables that set up the framework and its extensions, which will be described in the final part of this section, is summarised in the following list (references to temporal properties removed for easier reading):

- C: cell index on the BVM occupancy grid given by the 3D coordinates of its "far corner";
- Z: generic designation for either visual or auditory sensor measurements;
- O_C: binary value signalling the fact that a cell C is either empty or occupied by an object;
- V_C: discrete variable indicating instantaneous local motion vector for objects occupying cell C;
- G: fixation point for next gaze-shift in log-spherical coordinates;
- U_C: entropy gradient-based variable ranging from 0 to 1, signalling the potential interest (i.e. 0 and 1 meaning minimally and maximally interesting, respectively) of cell C as future focus of attention given the uncertainty on its current state given by (O_C, V_C), thus promoting an active exploration behaviour;
- S_C^i : binary value describing the ith of N sensory saliency of cell C;

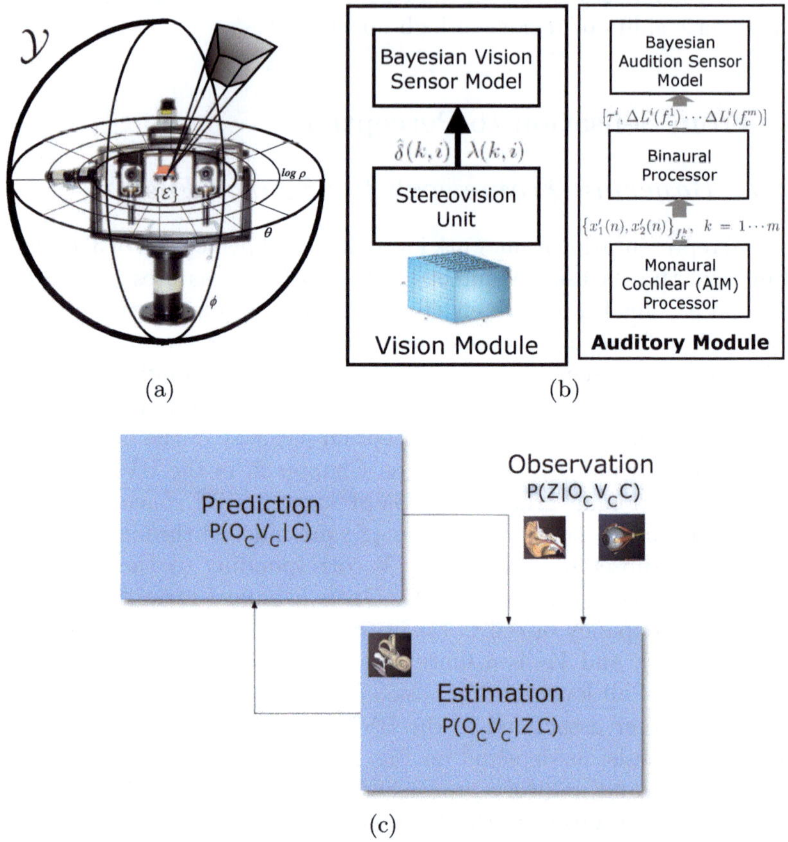

Fig. 8.2. Multimodal perception framework details. (a) The Bayesian Volumetric Map (BVM) referred to the egocentric coordinate frame of the robotic active perception system; (b) BVM sensor models; (c) BVM Bayesian Occupancy Filter.

- $Q^i_C = P([S^i_C = 1]|Z^i\,C)$: probability of a perceptually salient object occupying cell C;
- R_C: inhibition level for cell C as a possible future focus of attention modelling the Inhibition of Return behaviour, ranging from no inhibition (0) to full inhibition (1).

By restricting egomotion to rotations around the egocentric axes, vestibular sensing, together with the encoders of the motors of the robotic head (i.e. proprioception), yield measurements of angular velocity and position which can then be easily used to manipulate the BVM, which is, by definition, in spherical coordinates [19]. In this case, the most effective solution for integration is to perform the equivalent index shift. This process is described by

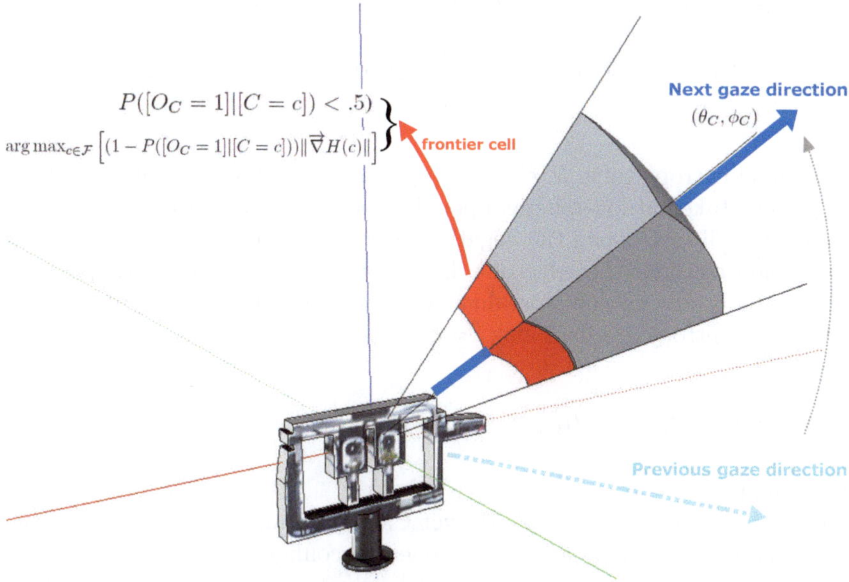

Fig. 8.3. Illustration of the entropy-based active exploration process using the Bayesian Volumetric Map. The result of applying the algorithm steps described in the main text is depicted. When there exists more than one maximum for $(1 - P([O_C = 1]|[C = c]))\|\vec{\nabla} H(c)\|$, the frontier cell corresponding to the direction closest to the current heading is chosen, so as to ensure minimum gaze shift rotation effort.

redefining C: $C \in \mathcal{Y}$ *indexes a cell in the BVM by its far corner, defined as* $C = (\log_b \rho_{max}, \theta_{max} + \theta_{inertial}, \phi_{max} + \phi_{inertial}) \in \mathcal{Y}$.

8.2.2 Extending the Update Model

The BVM is extendible in such a way that other properties, characterised by additional random variables and corresponding probabilities might be represented. To this end, other than the already implemented occupancy and local motion properties O_C and V_C, additional properties were implemented by augmenting the hierarchy of operators through Bayesian subprogramming (see Chapter 3).

One such property that we propose to model uses the knowledge from the BVM to determine gaze shift fixation sites. More precisely, it elicits gaze shifts towards locations of high entropy/uncertainty based on the rationale conveyed by an additional variable that quantifies the uncertainty-based interest of a cell on the BVM, thus promoting entropy-based active exploration.

Information in the BVM is stored as the *probability of each cell being in a certain state*, defined as $P(V_c\, O_c | z\, c)$. The state of each cell thus belongs to the state-space $\mathcal{O} \times \mathcal{V}$. The *joint entropy* of the random variables V_C and O_C

that compose the state of each BVM cell $[C = c]$ is defined as follows:

$$H(c) \equiv H(V_c, O_c) = - \sum_{\substack{o_c \in \mathcal{O} \\ v_c \in \mathcal{V}}} P(v_c\, o_c | z\, c) \log P(v_c\, o_c | z\, c) \tag{8.1}$$

The joint entropy value $H(c)$ is a sample of a continuous joint entropy field $H : \mathcal{Y} \rightarrow \mathbb{R}$, taken at log-spherical positions $[C = c] \in \mathcal{Y}$. Let $c_{\alpha-}$ denote the contiguous cell to C along the negative direction of the generic log-spherical axis α, and consider the edge of cells to be of unit length in log-spherical space, without any loss of generality. A reasonable first order approximation to the joint entropy gradient at $[C = c]$ would be

$$\vec{\nabla} H(c) \approx [H(c) - H(c_{\rho-}), H(c) - H(c_{\theta-}), H(c) - H(c_{\phi-})]^T \tag{8.2}$$

with magnitude $\|\vec{\nabla} H(c)\|$.

A great advantage of the BVM over Cartesian implementations of occupancy maps is the fact that the log-spherical configuration avoids the need for time-consuming ray-casting techniques when computing a gaze direction for active exploration, since the log-spherical space is already defined based on directions (θ, ϕ). Hence, the active exploration algorithm is simplified to the completion of the following steps (see Fig. 8.3):

1. Find the last non-occluded, close-to-empty (i.e. $P([O_C = 1]|[C = c]) < .5$) cell for the whole span of directions $(\theta_{\max}, \phi_{\max})$ in the BVM – these are considered to be the so-called *frontier cells* as defined on [31]; the set of all frontier cells will be denoted here as $\mathcal{F} \subset \mathcal{Y}$.
2. Compute the joint entropy gradient for each of the frontier cells and select $c_s = \arg\max_{c \in \mathcal{F}} \left[(1 - P([O_C = 1]|[C = c])) \|\vec{\nabla} H(c)\| \right]$ as the best candidate cell to direct gaze to. In case there is more than one global maximum, choose the cell corresponding to the direction closest to the current heading, so as to deal with equiprobability, while simultaneously ensuring minimum gaze shift rotation effort.
3. Compute gaze direction as being (θ_C, ϕ_C), where θ_C and ϕ_C are the angles that bisect cell $[C = c_s]$ (i.e. which pass through the geometric centre of cell c_s).

Therefore, we introduce a new random variable U_C, which takes this algorithm and expresses it in a compact mathematical form:

$$U_C = \begin{cases} (1 - P([O_C = 1]|C)) \frac{\|\boldsymbol{\nabla} H(C)\|}{\max \|\boldsymbol{\nabla} H(C)\|} & C \in \mathcal{F}, \\ 0 & C \notin \mathcal{F}. \end{cases} \tag{8.3}$$

(a) Left camera snapshot of a male speaker, at -41° azimuth relatively to the Z axis, which defines the frontal heading respective to the IMPEP "neck".

(b) BVM results for binaural processing only. Interpretation, from left to right: 1) sound coming from speaker triggers an estimate for occupancy from the binaural sensor model, and a consecutive exploratory gaze shift at approximately 1.6 seconds; 2) At approximately 10 seconds, noise coming from the background introduces a false positive, that is never again removed from the map (i.e. no sound does not mean no object, only no audible sound-source).

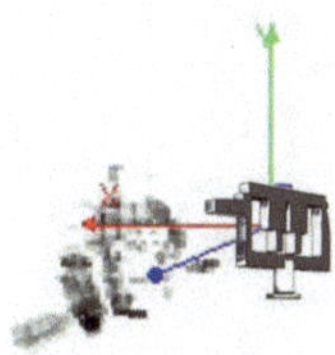

(c) BVM results for stereovision processing only. Although reconstruction detail is better than in (b), active exploration took approximately 15 seconds longer to start scanning the speaker's position in space, while using binaural processing the speaker was fixated a couple of seconds into the experiment.

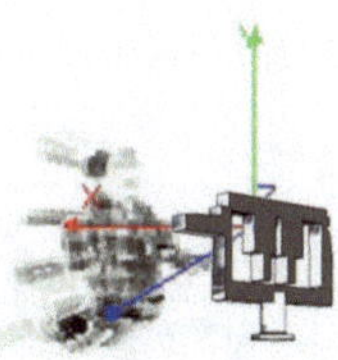

(d) BVM results for visuoauditory fusion. In this case, the advantages of both binaural (immediacy from panoramic scope) and stereovision (greater spatial resolution and the ability to clean empty regions in space) influence the final outcome of this particular instantiation of the BVM, taken at 1.5 seconds.

Fig. 8.4. Online results for the real-time prototype for multimodal perception of 3D structure and motion using the BVM – three reenactments of a single speaker scenario. A scene consisting of a male speaker talking in a cluttered lab is observed by the IMPEP active perception system and processed online by the BVM Bayesian filter, using the active exploration heuristics described in the main text, in order to scan the surrounding environment. The blue arrow together with an oriented 3D sketch of the IMPEP perception system depicted in each map denote the current gaze orientation. All results depict frontal views, with Z pointing outward.

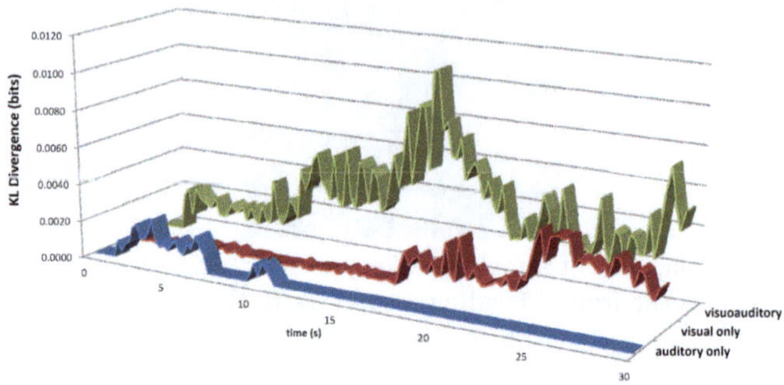

(a) Instantaneous average information gain for updated cells.

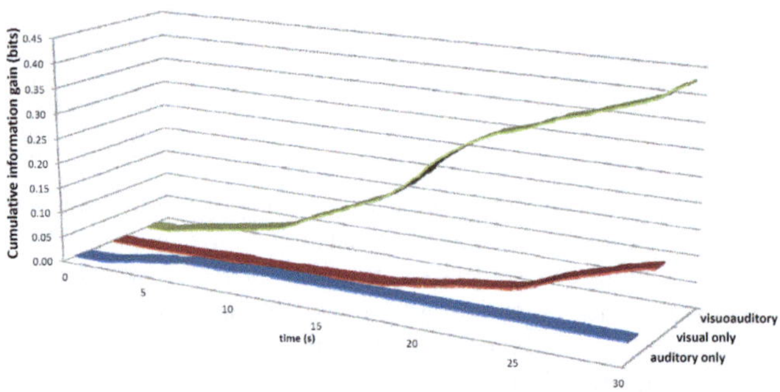

(b) Cumulative sum of average information gain for updated cells.

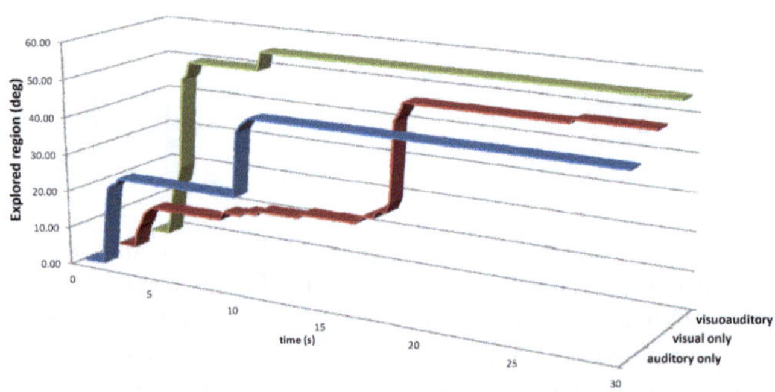

(c) Maximum angle covered by explored volumetric convex hull.

Fig. 8.5. Temporal evolution of average information gain (i.e. average Kullback-Liebler divergence for the full set of cells which were updated, either due to observations or propagation from prediction) and corresponding exploration span for the auditory-only, visual-only and visuoauditory versions of the single speaker scenario (see Fig. 8.4), for a 30 second period since the start of each experiment.

8.2.3 Experimental Evaluation of the Multisensory Active Exploration Behaviour Extension to the Update Model

In Fig. 8.4 a qualitative comparison is made between the outcome of using each sensory modality individually, and also with the result of multimodal fusion, using a single speaker scenario, showcasing the advantages of visuoauditory integration in the effective use of both the spatial precision of visual sensing, and the temporal precision and panoramic capabilities of auditory sensing. These representations were produced from screenshots of an online OpenGL-based viewer which would be running throughout the experiments. The parameters used for the BVM were as follows: $N = 10$, $\rho_{Min} = 1000\,\text{mm}$ and $\rho_{Max} = 2500\,\text{mm}$, $\theta \in [-180^o, 180^o]$, with $\Delta\theta = 1^o$, and $\phi \in [-90^o, 90^o]$, with $\Delta\phi = 2^o$, corresponding to $10 \times 360 \times 90 = 324,000$ cells, approximately delimiting the so-called "personal space" (the zone immediately surrounding the observer's head, generally within arm's reach and slightly beyond, within 2 m range [49]).

Fig. 8.5 presents a study based on information gain and exploration span, yielding a quantitative comparison of these advantages and capabilities, and demonstrating the superior results of visuoauditory fusion as compared to using each sensory modality separately.

8.3 Implementing the Action-Perception Loop

8.3.1 Bayesian Active Perception Hierarchy

To achieve our goal of designing Bayesian models for visuoauditory-driven saccade generation following human active perception behaviours, a hierarchical framework, inspired on what was proposed by Colas, Flacher, Tanner, Bessière, and Girard [10], has been developed and is presented in the following text.

We will specify three decision models: π_A, that implements entropy-based active exploration based on the BVM and the heuristics represented by equation 8.3, π_B, that uses entropy and saliency together for active perception, and finally π_C which adds a simple Inhibition of Return mechanism based on the fixation point of the previous time-step. In other words, each model incorporates its predecessor through Bayesian fusion, therefore constituting a model hierarchy – see Fig. 8.6.

The hierarchy is extensible in such a way that other properties characterised by additional random variables and corresponding probabilities might be represented, other than the already implemented occupancy and local motion properties of the BVM, by augmenting the hierarchy of operators through Bayesian subprogramming (Chapters 3 and 4). This ensures that the framework is *scalable*. On the other hand, the combination of these strategies to produce a coherent behaviour ensures that the framework is *emergent*.

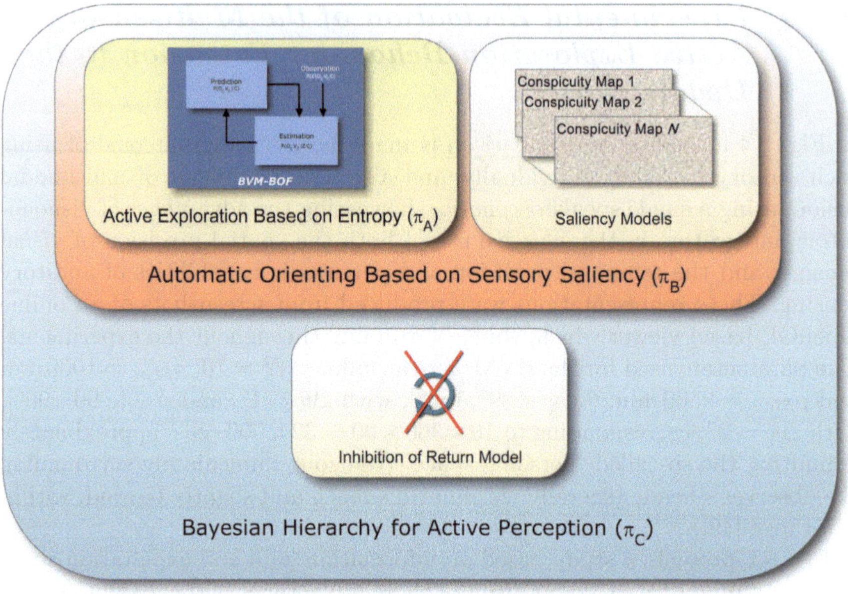

Fig. 8.6. Conceptual diagram for active perception model hierarchy

Furthermore, each model will infer a probability distribution on the next point of fixation for the next desired gaze shift represented by a random variable $G^t \in \mathcal{Y}$ at each time $t \in [1, t_{max}]$: $P(G^t | V^{1\to t}\, O^{1\to t}\, \pi_k)$, where $V^{1\to t} = \bigwedge_{t\in[1,t_{max}]} \bigwedge_C V_C^t$ and $O^{1\to t} = \bigwedge_{t\in[1,t_{max}]} \bigwedge_C O_C^t$ represent the conjunction of BVM local motion and occupancy estimate states for all cells $C \in \mathcal{Y}$, from system startup up until current time-instant t.

The first model we propose uses the knowledge from the BVM layer to determine gaze shift fixation points. More precisely, it tends to look towards locations of high entropy/uncertainty. Its likelihood is based on the rationale conveyed by the additional variable U_C, defined earlier.

The Bayesian Program for this model is presented on Fig. 8.7. The dependency of the uncertainty measure variable U_C^t – equation (8.3) – on the BVM states $(V^{1\to t}, O^{1\to t})$ are implicitly stated by definition, thus, with this model, the distribution on the point of fixation of the next desired gaze shift can be computed using the following expression

$$
\begin{aligned}
P(G^t | V^{1\to t}\, O^{1\to t}\, \pi_A) &= P(G^t | U^t\, \pi_A) \\
&\propto \prod_C P(U_C^t | G^t\, \pi_A)
\end{aligned}
\tag{8.4}
$$

The second model is based on sensor models that relate sensor measurements $Z_j^{i,t}$ with $i = 1..N$ independent sensory properties of saliency ($j = 1..M_i$ total independent measurements for each saliency property),

Program / Description / Specification

Relevant variables:

O_C^t: binary value describing the occupancy of cell C at time t,
$[O_C = 1]$ if cell C is occupied by an object,
$[O_C = 0]$ otherwise (related variable: $O^t = \bigwedge_C O_C^t$);

V_C^t: velocity of cell C at time t, discretised into $n+1$ possible cases
$\in \mathcal{V} \equiv \{v_0, \cdots, v_n\}$ (related variable: $V^t = \bigwedge_C V_C^t$);

$G^t \equiv (\log_b \rho_{\max}, \theta_{\max}, \phi_{\max}) \in \mathcal{Y}$: fixation point for next gaze-shift, computed at time t, with related variable $G = \bigwedge_{t \in [1, t_{max}]} G^t$;

$U_C^t \equiv f(V^{1\to t}, O^{1\to t}) \in [0,1]$: joint entropy gradient-based variable at time t, with related variables $U^t = \bigwedge_C U_C^t$ and $U = \bigwedge_{t \in [1, t_{max}]} U^t$
(see equation (8.3): close to 1 when uncertainty is high and C is a frontier cell, and $U_C \to 0$ when uncertainty is low or C is not a frontier cell).

Decomposition:

$$P(G\,U|\pi_A) = \prod_{t=1}^{t_{max}} \left[P(G^t|\pi_A) \prod_C P(U_C^t|G^t\,\pi_A) \right]$$

Parametric forms:

$P(G^t|\pi_A)$: uniform prior;

$P(U_C^t|G^t\,\pi_A)$: is a beta distribution $\mathcal{B}(\alpha_U, \beta_U)$ for $[G^t = C]$ that expresses that, for a given point of fixation proposal for the next gaze shift, U_C^t is more likely near 1, or a uniform distribution on U_C^t for $[G^t \neq C]$.

Identification:

Empirical values for free parameters α_U and β_U.

Questions:

$P(G^t|V^{1\to t}\,O^{1\to t}\,\pi_A) = P(G^t|U^t\,\pi_A)$

Fig. 8.7. Bayesian Program for entropy-based active exploration model π_A

represented by the set of binary random variables $S_C^{i,t}$ (equalling 0 when the cell is non-salient and 1 when salient) corresponding to each cell C. In other words, these sensor models are generically notated as $P(Z^t|S_C^{i,t}\,V_C^t\,O_C^t\,\pi_C)$, indiscriminately of what the specific sensory saliency property $S^{i,t} = \bigwedge_C S_C^{i,t}$ might represent.

The Bayesian Program for model π_B is presented on Fig. 8.8. With this model, the distribution on the point of fixation of the next desired gaze shift can be computed using the following expression

$$\begin{aligned} &P(G^t|V^{1\to t}\,O^{1\to t}\,S^t\,\pi_B) \propto \\ &\quad P(G^t|V^{1\to t}\,O^{1\to t}\pi_A) \prod_C \left[\prod_{i=1}^{N} P(Q_C^{i,t}|G^t\,\pi_B) \right] \end{aligned} \tag{8.5}$$

Program

Description

Specification

Relevant variables:

O_C^t: binary value describing the occupancy of cell C at time t, $[O_C = 1]$ if cell C is occupied by an object, $[O_C = 0]$ otherwise (related variables: $O = \bigwedge_{t \in [1, t_{max}]} O_C^t$ and $O^t = \bigwedge_C O_C^t$);

V_C^t: velocity of cell C at time t, discretised into $n+1$ possible cases $\in \mathcal{V} \equiv \{v_0, \cdots, v_n\}$ (related variables: $V = \bigwedge_{t \in [1, t_{max}]} V_C^t$ and $V^t = \bigwedge_C V_C^t$);

$G^t \equiv (\log_b \rho_{\max}, \theta_{\max}, \phi_{\max}) \in \mathcal{Y}$: fixation point for next gaze-shift, computed at time t, with related variable $G = \bigwedge_{t \in [1, t_{max}]} G^t$;

$S_C^{i,t}$: binary value describing the ith of N sensory saliency properties of cell C at time t, $[S_C^{i,t} = 0]$ when non-salient and $[S_C^{i,t} = 1]$ when salient (related variables: $S^i = \bigwedge_{t \in [1, t_{max}]} S^{i,t}$, $S^t = \bigwedge_{i=1}^{N} S^{i,t}$ and $S = \bigwedge_{i=1}^{N} S^i$);

$Z_j^{i,t} \in \mathcal{Z}$: sensor measurements at time t ($j = 1..M_i$ total independent measurements for each saliency property at time t, $S^{i,t}$) (related variables: $Z^i = \bigwedge_{t \in [0, t_{max}]} Z^{i,t}$ and $Z = \bigwedge_{i=1}^{N} Z^i$);

$Q_C^{i,t} = P([S_C^{i,t} = 1] \mid Z_j^{i,t} C) \in [0, 1]$: probability of a perceptually salient object occupying cell C (related variables: $Q^i = \bigwedge_{t \in [1, t_{max}]} Q^{i,t}$, $Q^t = \bigwedge_{i=1}^{N} Q^{i,t}$ and $Q = \bigwedge_{i=1}^{N} Q^i$).

Decomposition:

$$P(G\, Q \mid \pi_B) = \prod_{t=1}^{t_{max}} \left\{ P(G^t \mid \pi_B) \prod_C \left[\prod_{i=1}^{N} P(Q_C^{i,t} \mid G^t\, \pi_B) \right] \right\}$$

Parametric forms:

$P(G^t \mid \pi_B) \equiv P(G^t \mid V^{1 \to t}\, O^{1 \to t}\, \pi_A)$ is the prior taken from the result of the model of Figure 8.7; $P(Q_C^{i,t} \mid G^t\, \pi_B)$ is a beta distribution $\mathcal{B}(\alpha_Q, \beta_Q)$ for $[G^t = C]$ that expresses that, for a given point of fixation proposal for the next gaze shift, $Q_C^{i,t}$ is more likely near 1, or a uniform distribution on $Q_C^{i,t}$ for $[G^t \neq C]$.

Identification:

Empirical values for free parameters α_Q and β_Q.

Questions:

$P(G^t \mid V^{1 \to t}\, O^{1 \to t}\, S^t\, \pi_B) = P(G^t \mid Q^t\, \pi_B)$

Fig. 8.8. Bayesian Program for automatic orienting based on sensory saliency model π_B

In short, this model is the product between the prior on gaze shifts due to entropy-based active exploration and each distribution on the sensory-salient cells. This expression shows that the model is attracted towards both salient cells (without necessarily looking at one in particular, as the balance between the distributions on salient cells can lead to a peak in some weighted sum of their locations) *and* locations of high uncertainty when sensory saliency is not preponderant enough (i.e. this process is called *weighting*, as opposed to *switching*, in which these behaviours would be mutually exclusive – see Colas et al. [7]; Ferreira and Castelo-Branco [23]).

Program / Description / Specification

Relevant variables:

O_C^t: binary value describing the occupancy of cell C at time t, $[O_C = 1]$ if cell C is occupied by an object, $[O_C = 0]$ otherwise (related variables: $O = \bigwedge_{t\in[1,t_{max}]} O_C^t$ and $O^t = \bigwedge_C O_C^t$);

V_C^t: velocity of cell C at time t, discretised into $n+1$ possible cases $\in \mathcal{V} \equiv \{v_0, \cdots, v_n\}$ (related variables: $V = \bigwedge_{t\in[1,t_{max}]} V_C^t$ and $V^t = \bigwedge_C V_C^t$);

$R_C^t \equiv f(G^{t-1}) \in [0,1]$: inhibition level for cell C modelling the Inhibition of Return behaviour (see below), ranging from no inhibition (0) to full inhibition (1) (related variables: $R^t = \bigwedge_C R_C^t$ and $R = \bigwedge_{t\in[1,t_{max}]} R^t$);

$G^t \equiv (\log_b \rho_{\max}, \theta_{\max}, \phi_{\max}) \in \mathcal{Y}$: fixation point for next gaze-shift, computed at time t, with related variable $G = \bigwedge_{t\in[1,t_{max}]} G^t$;

$S_C^{i,t}$: binary value describing the ith of N sensory saliency properties of cell C at time t, $[S_C^{i,t} = 0]$ when non-salient and $[S_C^{i,t} = 1]$ when salient (related variables: $S^i = \bigwedge_{t\in[1,t_{max}]} S^{i,t}$, $S^t = \bigwedge_{i=1}^{N} S^{i,t}$ and $S = \bigwedge_{i=1}^{N} S^i$);

$Z_j^{i,t} \in \mathcal{Z}$: sensor measurements at time t ($j = 1..M_i$ total independent measurements for each saliency property at time t, $S^{i,t}$) (related variables: $Z^i = \bigwedge_{t\in[0,t_{max}]} Z^{i,t}$ and $Z = \bigwedge_{i=1}^{N} Z^i$);

$Q_C^{i,t} = P([S_C^{i,t} = 1] | Z_j^{i,t}\, C) \in [0,1]$: probability of a perceptually salient object occupying cell C (related variables: $Q^i = \bigwedge_{t\in[1,t_{max}]} Q^{i,t}$, $Q^t = \bigwedge_{i=1}^{N} Q^{i,t}$ and $Q = \bigwedge_{i=1}^{N} Q^i$).

Decomposition:

$$P(G\, R | \pi_C) = \prod_{t=1}^{t_{max}} \left\{ P(G^t | \pi_C) \prod_C \left[P(R_C^t | G^t\, \pi_C) \right] \right\}$$

Parametric forms:

$P(R_C^t | G^t\, \pi_C)$: is a beta distribution $\mathcal{B}(\alpha_R, \beta_R)$ for $[G^t = C]$ modelling the Inhibition of Return behaviour (see main text) that expresses that, for a given point of fixation proposal for the next gaze shift, R_C^t is more likely to be 0, and a uniform distribution for $[G^t \neq C]$.

$P(G^t | \pi_C) \equiv P(G^t | V^{1\to t}\, O^{1\to t}\, S^t\, \pi_B)$ is the prior taken from the result of the model of Figure 8.8;

Identification:

Empirical values for free parameters α_R and β_R.

Questions:

$P(G^t | V^{1\to t}\, O^{1\to t}\, S^t\, G^{t-1} \pi_C) = P(G^t | R^t\, \pi_C)$

Fig. 8.9. Bayesian Program for full active perception model π_C

The Bayesian Program for the third and final model π_C, which defines the full active perception hierarchy by adding an implementation the Inhibition of Return (IoR) mechanism, is presented on Fig. 8.9. With this model, the distribution on the point of fixation of the next desired gaze shift can be computed using the following expression

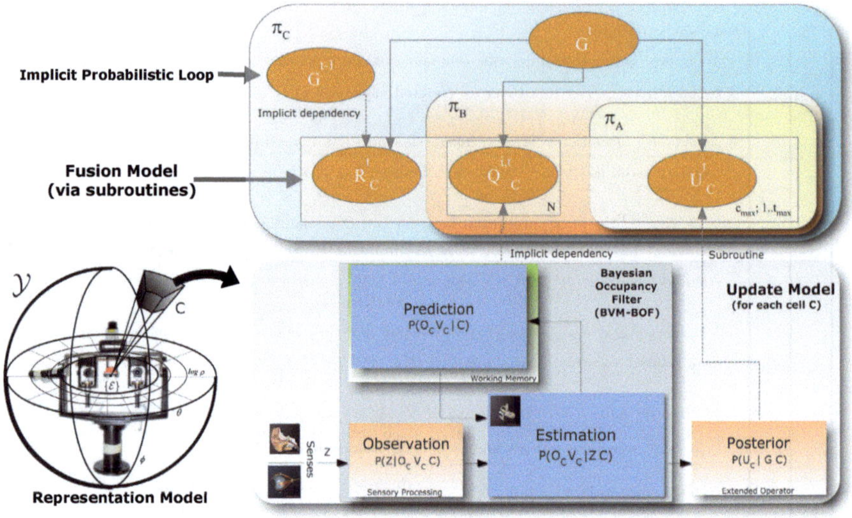

Fig. 8.10. Graphical representation of the hierarchical framework for active perception. Bottom half: Update and Representation Models for the BVM-BOF framework, extended by the entropy gradient-based operator. Upper half: Bayesian network summarising the models presented in this text, using the plates notation (introduced in Chapter 3). As can be seen, emergent behaviour results from a probabilistic fusion model implemented through a sequence of Bayesian Programming subroutines and an implicit loop that ensures the dynamic behaviour of the framework.

$$P(G^t|V^{1\to t}\, O^{1\to t}\, S^t\, G^{t-1}\, \pi_C) \propto \\ P(G^t|V^{1\to t}\, O^{1\to t}\, S^t\, \pi_B) \prod_C \left[P(R^t_C|G^t\, \pi_C)\right] \tag{8.6}$$

In conclusion, the full hierarchy, represented graphically in Fig. 8.10, is defined as the product between the prior on gaze shifts due to entropy-based active exploration and each distribution on the sensory-salient cells, while avoiding the fixation site computed on the previous time step through the IoR process, implemented by the last factor in the product. The parameters used for each distribution in this product, which define the relative importance of each level of the hierarchy and of each sensory saliency property, may be introduced directly by the programmer (like a genetic imprint) or manipulated "on the fly", which in turn allows for goal-dependent behaviour implementation (i.e. top-down influences), therefore ensuring that the framework is *adaptive*.

8.3.2 Parametrising the Models to Enact Complex Behaviour

Saliency properties from a generic visual cue, or, in other words, the conspicuity maps given by the BVM extended operators $Q^{i,t}_C = P([S^{i,t}_C = 1]|Z^{i,t}_j\, C) \in [0,1]$, were implemented in two steps:

1. A single-channel image with values varying between 0 and 1 is taken directly from visual cues taken from the right camera of the stereovision setup (thus simulating a dominant eye), either by directly normalising traditional dense conspicuity maps as defined by Itti et al. [47], or by generating a conspicuity map by forming Gaussian distributions with specific standard deviations centred on individual points of interest on the right camera image, for example in the case of sparse feature extractors such as face detection algorithms.
2. The saliency values from each pixel in the conspicuity map for which a disparity was estimated by the stereovision module are then projected on the log-spherical configuration through projection lines spanning the corresponding (θ, ϕ) angles – if two or more different saliency values are projected throughout the same direction, only the highest saliency value is used. These values are thus considered as soft evidence regarding $S_C^{i,t}$, therefore yielding $Q_C^{i,t}$.

The specific properties used in this work (although any visual saliency property would have been usable by applying the two steps described above) were optical flow magnitude taken from the result of using the CUDA implementation of the "Bayesian Multi-scale Differential Optical Flow" algorithm of Simoncelli [46] by [15], and face detection using the Haar-like features implementation of the OpenCV library.

The auditory saliency property used in this work was directly implemented from the $P([S_C = 1] | Z\, C)$ question solved by the Bayesian model of binaural perception.

The Inhibition of Return mechanism used in this work was implemented by assigning values on a log-spherical data structure corresponding to R_C^t ranging from 1 to values close to 0 depending on the distance in $\mathcal{Y}$ between G^{t-1} and each C, denoted d_{IoR}, through the following expression

$$R_C^t \equiv f(G^{t-1}) = \left(\frac{1}{2}\right)^{d_{IoR}} \tag{8.7}$$

The parameters of the Beta distributions defined on the Bayesian Programs of Figs. 8.7, 8.8 and 8.9 will, in general, function as relative importance weights for each behaviour in the fusion process. However, in two extreme cases, theses parameters will serve as a switch: $\alpha = 1, \beta = 1$ will result in degenerating the corresponding beta distribution into a uniform distribution, hence switching off the respective behaviour, while $\alpha >> \beta$ or $\beta >> \alpha$ will degenerate the Beta distribution into a Dirac delta function, hence serving as a mutually exclusive switch, "numerically deactivating" all other behaviours.

A set of parameters was chosen for initial values in order to attain the beta distributions presented on Fig. 8.11. These preprogrammed parameters define the genetic imprint of preliminary knowledge that establishes the baseline hierarchy of the set of active perception behaviours; these

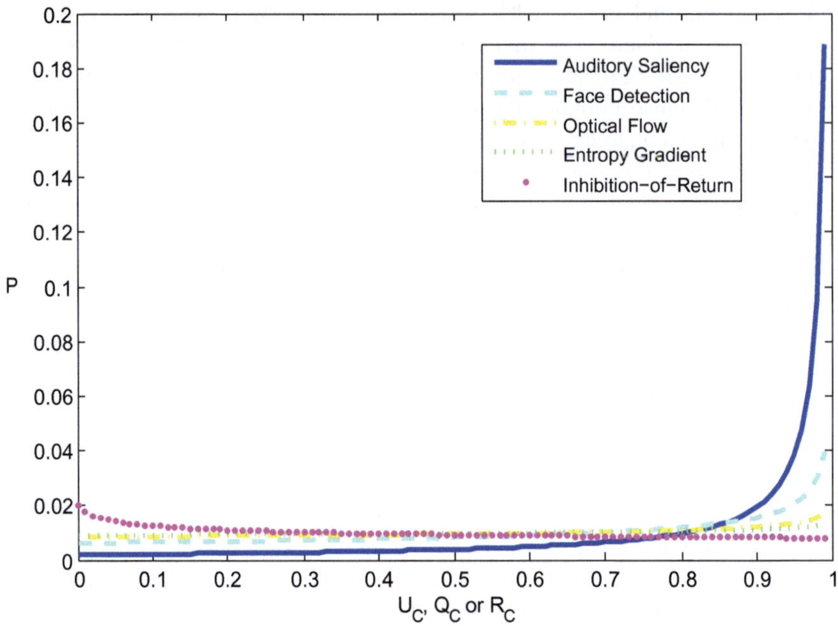

Fig. 8.11. Beta distributions of the active perception hierarchy using the baseline choice for parameters. Corresponding parameters are $\alpha_U = 1$ and $\beta_U = 0.92$ for active exploration, $\alpha_Q = 1$ and $\beta_Q = 0.01$ for auditory saliency, $\alpha_Q = 1$ and $\beta_Q = 0.6$ for face detection saliency, $\alpha_Q = 1$ and $\beta_Q = 0.85$ for optical flow magnitude saliency, and $\alpha_R = 0.8$ and $\beta_R = 1$ for Inhibition of Return.

parameters are changeable "on the fly" through sliders on the graphical user interface of the implementation software, thus simulating top-down influences on behaviour prioritisation (i.e. the *adaptivity* property). The influence of the relative weights imposed by this choice of parameters will be discussed on the Results section.

The fixation point for the next time instant G^t is obtained by substituting equations (8.4) and (8.5) consecutively into (8.6), and computing $G^t = \arg\max_C P([G^t = C]|V^{1\to t}\, O^{1\to t}\, S^t\, G^{t-1}\, \pi_C)$, knowing that

$$
\begin{aligned}
&P([G^t = C]|V^{1\to t}\, O^{1\to t}\, S^t\, G^{t-1}\, \pi_C) \propto \\
&\quad P(U_C^t|[G^t = C]\,\pi_A) \prod_{i=1}^{N} \left[P(Q_C^{i,t}|[G^t = C]\,\pi_B)\right] P(R_C^t|[G^t = C]\,\pi_C)
\end{aligned}
\tag{8.8}
$$

by factoring out the effect of the uniform distributions corresponding to considering $[G^t \neq C]$.

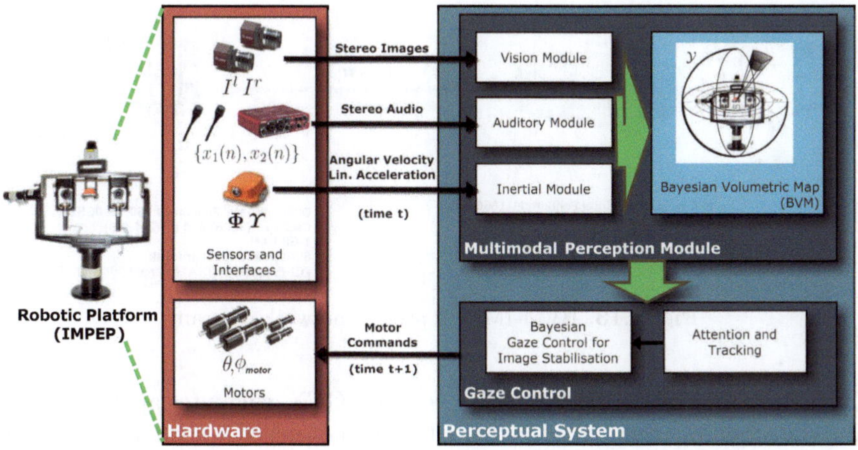

Fig. 8.12. Implementation diagram for the BVM-IMPEP multimodal perception framework

8.3.3 System Overview and Implementation

The BVM-IMPEP framework, of which an implementation diagram is presented on Fig. 8.12, was realised as follows:

- **Vision sensor system:** with the OpenCV toolbox and the implementation by Gallup [14] of a basic binocular stereo algorithm on GPU using NVIDIA's general purpose parallel computing architecture CUDA[2]. The algorithm reportedly runs at 40 Hz on 640×480 images while detecting 50 different levels of disparity, computing left and right disparity maps and performing left-right consistency validation (which in our adaptation is used to produce the stereovision confidence maps).
- **Binaural sensor system:** Using an adaptation of the real-time software kindly made available by the Speech and Hearing Group at the University of Shefield [25] to implement binaural cue analysis as described in Chapter 2.
- **Bayesian Volumetric Map, Bayesian sensor models and active exploration:** using our proprietary, parallel processing, GPU implementation developed with CUDA.

The BVM-IMPEP system is composed of a local Ethernet network comprised of two PCs communicating and synchronising via Carmen messaging (Carmen Robot Navigation Toolkit – `http://carmen.sourceforge.net/home.html` – an open-source collection of software for mobile robot control sponsored by DARPA's MARS Program), one for all the sensory and BVM framework processing (including

[2] Refer to Appendix A for an introduction to this technology.

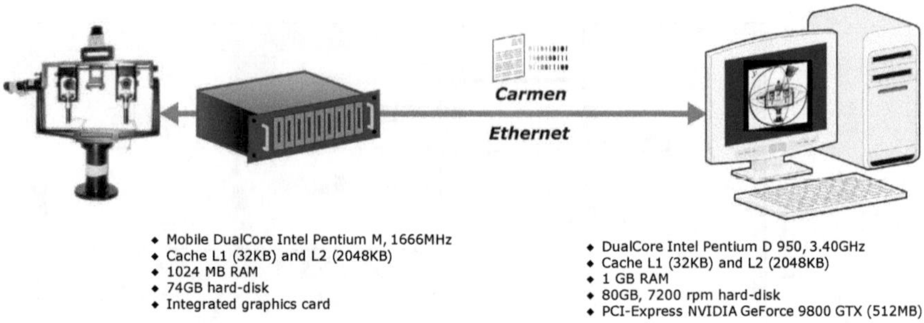

Fig. 8.13. BVM-IMPEP system network diagram

CUDA processing on a NVIDIA GeForce 9800 GTX, compute capability 1.1), and the other for controlling the IMPEP head motors, designed for portability (i.e. low-consumption and light-weight) in order to be mounted on mobile robotic platforms in the future – see Fig. 8.13. Both are equipped with Ubuntu Linux v9.04.

The activity diagram for the BVM Bayesian framework is presented on Fig. 8.14, depicting an inference step corresponding to time t and respective timeline. In the following lines, our GPU implementation of the BVM algorithms developed with NVIDIA's CUDA that execute this timeline will be described in more detail – for more on CUDA and a brief overview of the implementation of perception algorithms using GPU Computing, please refer to Appendix A.

The BVM filter, which comprises the processing lane on the right of Fig. 8.14, launches kernels based on a single three-dimensional grid corresponding to the log-spherical configuration – see Fig. 8.15. In fact, both input matrices (i.e. observations and previous system state matrices) and output matrices (i.e. current state matrices) have the same indexing system. Blocks on this grid were arranged in such a way that their 2D indices would coincide with azimuth θ and elevation ϕ indices on the grid, assuming that the full N-depth of the log-distance index is always copied to shared memory.

By trial-and-error we arrived at the conclusion that block size was limited by shared memory resources to $5 \times 5 \times N$ for $N \leq 10$ and $3 \times 3 \times N$ for $N = 11$, which would therefore be the top limit for depth using this rationale of a single grid for the whole BVM space. In fact, for $N < 11$, there were 250 threads and 8000 bytes of shared memory per block, thus limiting the maximum number of blocks per multiprocessor to 2 for the compute capability 1.1 of the GeForce 9800 GTX; for $N = 11$, on the other hand, there were 90 threads and 2880 bytes of shared memory per block, increasing the limit of blocks per multiprocessor to 5.

The flowchart for the BVM filter kernel is shown on Fig. 8.17(a).

The stererovision sensor model, which comprises the second processing lane from the left and the "Observation" box of Fig. 8.14, launches kernels

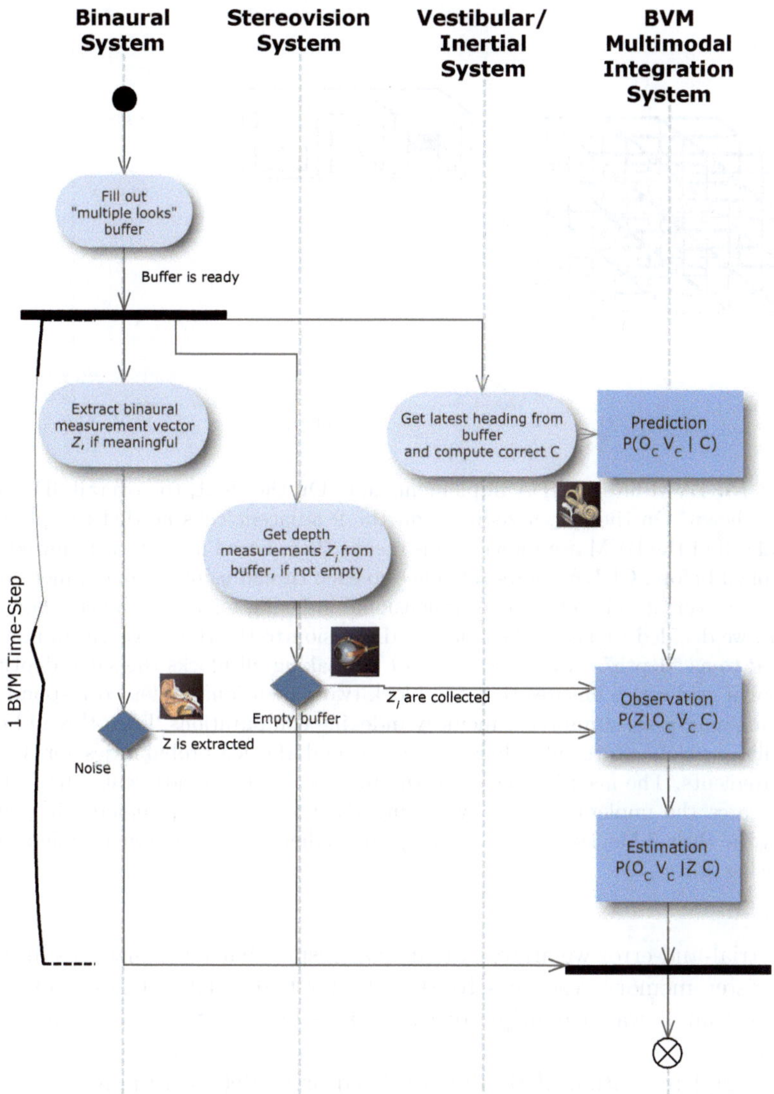

Fig. 8.14. Activity diagram for an inference time-step at time t. Each vertical lane represents a processing thread of the module labelled in the corresponding title. Maximum processing times (for $N = 10, \Delta\theta = 1°, \Delta\phi = 2°$) are also presented in the timeline for reference.

based on two-dimensional grids corresponding to image configuration – see Fig. 8.16. In fact, its input matrices (left and right images, and disparity and confidence maps) have the same indexing system, while its output matrices (visual observation matrices) have the same indexing as the BVM grid of Fig. 8.15.

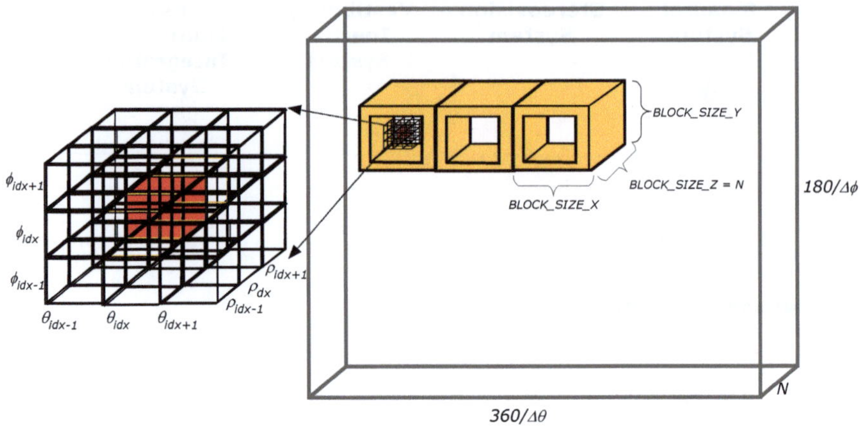

Fig. 8.15. BVM filter CUDA implementation. On the right, the overall 3D BVM grid is shown. On the left, a zoom in on the 9 adjacent cells needed to update a central cell of the BVM are shown – this means that shared memory is required. As mentioned before, CUDA allows reference to each thread using a three-dimensional index; however, it only allows two-dimensional indexing for thread blocks. For this reason, we decided to assign the smallest dimension to the third axis (from now on referred to as "depth" – with size N), and by making all blocks the same depth as the global grid – this ensures that the block two-dimensional index corresponds to the remaining axes, simplifying memory indexing computations. Each thread loads its cell's previous state into shared memory and the log-probabilities for sensor measurements. The need for access to the previous states of adjacent cells further complicates the implementation by forcing the use of *aprons*, depicted in yellow within the thread blocks (see Fig. 8.17(a) for further details on kernel implementation using aprons).

By trial-and-error we arrived at the conclusion that block size was limited by register memory resources to 16×16 for 640×480 images. This also ensured that it was a multiple of the warp size so as to achieve maximum efficiency.

The implementation of the binaural sensor model, corresponding to the processing lane on the left and the "Observation" box of Fig. 8.14, contrastingly, is very simple – a vector of binaural readings is used as an input and a grid as shown on Fig. 8.15, but without resorting to aprons (i.e. shared memory; see Figs. 8.15 and 8.17(a) for a detailed explanation of this notion), was used to update sensor model measurement data structures analogous to those of the stereovision sensor model, by referring to a lookup table with normal distribution parameters taken from the auditory system calibration procedure (see Chapter 2).

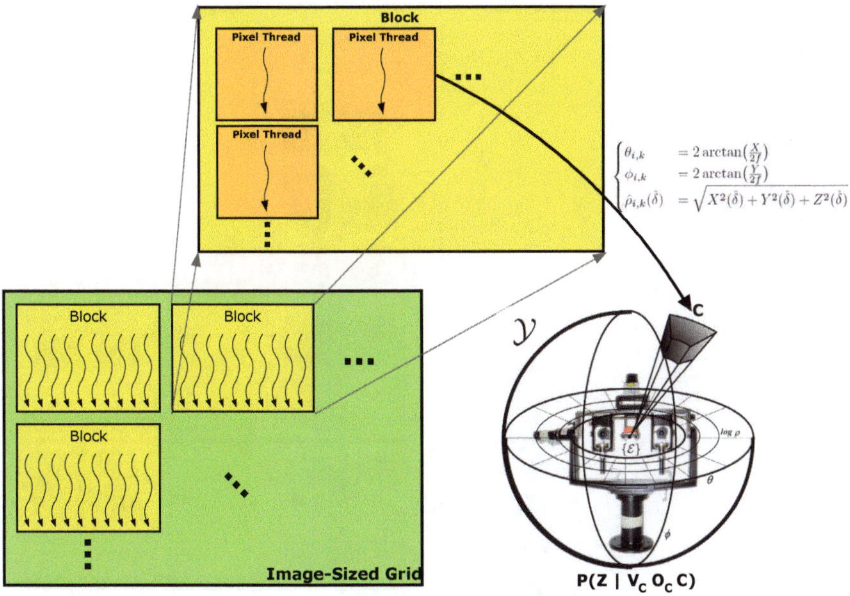

Fig. 8.16. Stereovision sensor model CUDA implementation. Each thread independently processes one pixel of the egocentric-referred depth map and confidence images (no use of shared memory required), computes the corresponding cell C on the BVM log-spherical spatial configuration using the equation shown, and updates two data structures in global device memory with that configuration storing log-probabilities corresponding to $P(Z|[O_C = 1]\, C)$ and $P(Z|[O_C = 0]\, C)$ (independent of velocity V_C), respectively. The update is performed using atomic summation operations provided by CUDA compute capability 1.1 and higher [26]. Atomic operations are needed due to the many-to-one correspondence between pixels and cells on the BVM; however, the order of summation is, obviously, non-important. Finally, since all atomic operations except "exchange" only accept integers as arguments, log-probabilities are converted from to floating-point to integer through a truncated multiplication by 10^n, with n corresponding to the desired precision (in our implementation, we used $n = 4$).

When there are visual and binaural measurements available simultaneously, two CUDA streams[3] are created (i.e. forked), one for each sensor model, and then destroyed (i.e. merged).

The active exploration algorithm was implemented resorting to CUDA atomic operations, global memory and four consecutive kernels in a sequential CUDA stream. This implementation is detailed on Fig. 8.17(b).

To avoid the adverse effects of motion blur on stereoscopic measurements, a strategy similar to what is adopted by the human brain is implemented for

[3] CUDA *streams* are concurrent lanes of execution that allow parallel execution of multiple kernels on the GPU.

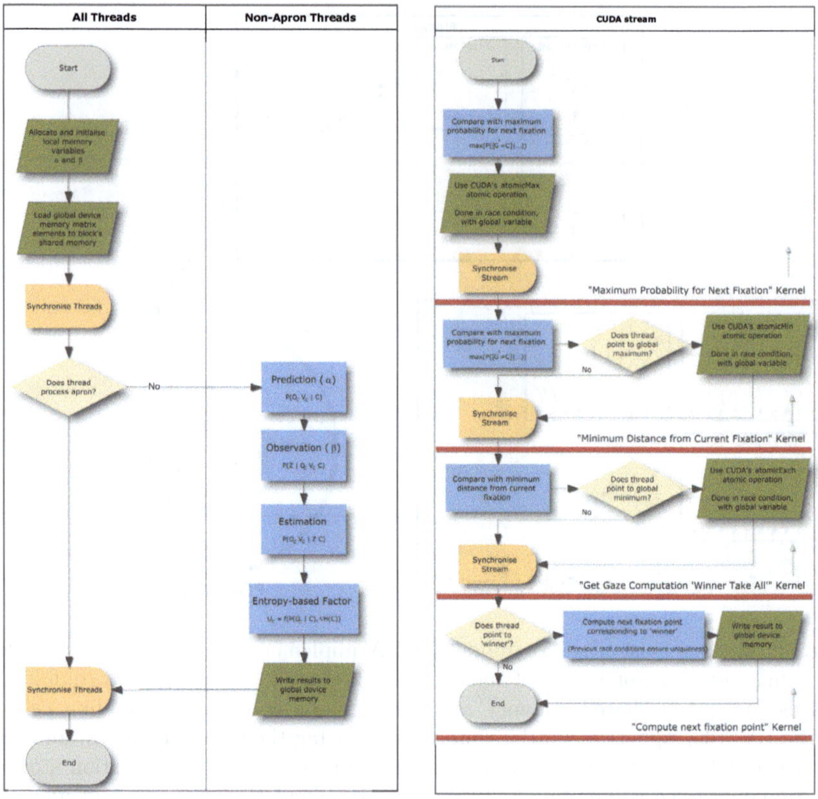

(a) BVM filter CUDA kernel flowchart. *Aprons* are the limiting cells of the block, to which correspond threads that cannot access adjacent states, and therefore with the sole mission of loading their respective states into shared memory – thus, blocks must overlap as their indices change, so that all cells have the chance to be non-apron. After all threads, apron or non-apron, load their respective previous states into shared memory, all non-apron threads then perform Bayesian filter estimation and update the states, as depicted. The "Observation" box here denotes the computation of β by multiplying all available outputs from the stereovision and binaural Bayesian sensor models denoted as the "Observation" box of Fig. 8.14. Saliency variables (not shown) are estimated concurrently with the entropy-based factor.

(b) Active perception CUDA stream flowchart. Four consecutive kernels in a sequential CUDA stream were used to implement the active perception hierarchy. The division of the processing workload into separate kernels was necessary due to the fact that the only way to enforce synchronisation between all concurrent CUDA threads in a **grid** (as opposed to all threads in a **block**, which is only a subset of the former) is to wait for all kernels running on that grid to exit – this is only possible at CUDA stream level (see main text for the definition of CUDA stream). CUDA atomic operations (refer to Fig. 8.16 for more information) and global memory were used to pass on data from one kernel to the next without the need for additional memory operations.

Fig. 8.17. BVM CUDA implementation flowcharts

fixations and gaze shifts – fixation is accomplished by processing data coming from the stereovision system for a few hundred milliseconds [33; 43; 34], followed by a process similar to the so-called *saccadic suppresion*, in which the magnocellular visual pathway (mainly supplying data to the dorsal pathway, which we intend to model) is actively suppressed during saccades [52; 17]. In our parallel of this process, we simply halt gaze shift generation for a few iterations of the vision sensor model updates (thus simulating fixation), and then stop the updates during gaze shifts (thus simulating saccadic suppresion), without stopping the low-level processing of the stereovision system, which might be used in the future for other purposes.

The real-time implementation of all the processes of the framework was first subjected to performance testing for each individual module excluding the top model of the hierarchy, and reported in [5]. Processing times and rates for the sensory systems were as follows:

- **Stereovision unit** 15 Hz, including image grabbing and preprocessing (using CPU), stereovision processing itself (i.e. disparity and confidence map generation, using GPU), and postprocessing and numerical conditioning (using CPU).
- **Binaural processing unit** Maximum rate of 40 Hz and 20 to 70 ms latency (using CPU) for 44 KHz, 16-bit audio, with 16 frequency channels and 50 ms buffer for cue computation.
- **Inertial processing unit** 100 Hz using GPU.

Processing times for the BVM modules using the active exploration behaviour alone were measured, and are shown in Fig. 8.18. As can be seen, the system running only with active exploration runs from 6 to 10 Hz, depending on system parameters. This is ensured by forcing the main BVM thread to pause for each time-step when no visual measurement is available (e.g. during 40 ms for $N = 10, \Delta\phi = 2^{\circ}$ – see Fig. 8.14). This guarantees that BVM time-steps are regularly spaced, which is a very important requirement for correct implementation of prediction/dynamics, and also ensures that processing and memory resources are freed and unlocked regularly.

However, when augmenting the active perception hierarchy to work using all behaviours, the 15 Hz performance of the stereovision unit reported above was reduced to about 6 Hz, mainly as a consequence of the slow performance of the face detection algorithm. As a consequence, the full active perception system, implementing all behaviours, runs at about 5 Hz, for $N = 10$, $\Delta\theta = 1^{\circ}$, $\Delta\phi = 2^{\circ}$. Since the human saccade-generation system promotes fixation periods (i.e. time intervals between gaze shifts) of a few hundred milliseconds on average [33; 34], the overall rates achieved with our CUDA implementation back up the claim that our system does, in fact, achieve satisfactory real-time performance.

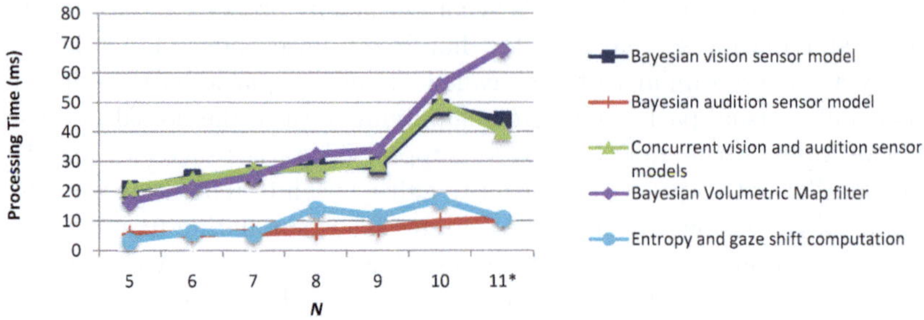

(a) Average processing times for $\Delta\phi = 2^{\circ}$.

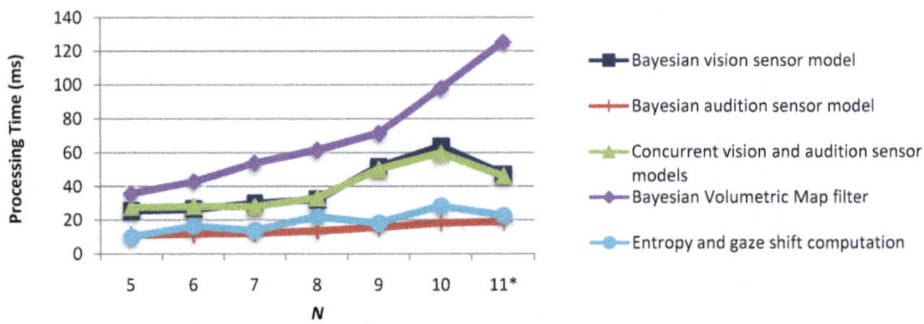

(b) Average processing times for $\Delta\phi = 1^{\circ}$.

Fig. 8.18. BVM framework average processing times. Both graphs are for $\Delta\theta = 1^{\circ}$, and show the average of processing times in ms for each activity depicted on Fig. 8.14, taken for a random set of 500 runs of each module in the processing of 5 dynamic real-world scenarios, with sensory horopter occupation varying roughly from 10 to 40% (although with no apparent effect on performance). These times are plotted against the number N of divisions in distance, which is the most crucial of system parameters (for $N > 11$, the GPU resources become depleted, and for $N < 5$ resolution arguably becomes unsatisfactory), and for two different reasonable resolutions in ϕ. Note that BVM filter performance degrades approximately exponentially with increasing resolution in distance, while the performance of all other activities degrades approximately linearly – the sole exception is the vision sensor model for $N = 11$, where it actually improves its performance. The reason for this is that the ratio of the effect of the influence of resolution on CUDA grid size *vs* the effect of the influence of resolution on the number of atomic operations required is reversed. (The * denotes that for $N = 11$ the block size is smaller for the BVM filter CUDA implementation – refer to main text for further details.)

Table 8.1. Summary table of experimental session planification

Session	Description	Objective
1	Implement all behaviours, using baseline priorities	Baseline demonstration, proof-of concept of emergence
2	Implement all behaviours, using swapped priorities	Proof-of-concept of adaptivity
3	Implement active exploration only	1st proof-of concept of scalability
4	Implement optical flow magnitude saliency only	2nd proof-of-concept of scalability
5	Implement optical flow Inhibition of Return only	3rd proof-of-concept of scalability

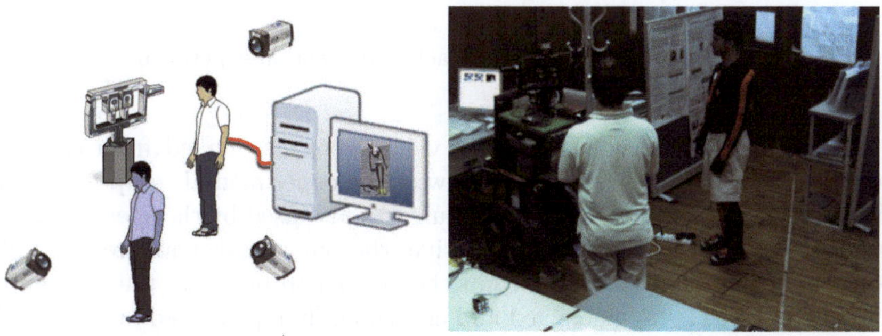

Fig. 8.19. Overview of the setup used in the experimental sessions testing the Bayesian hierarchical framework for multimodal active perception. The "IMPEP 2 and interlocutors" scenario, in which one of the interlocutors is wearing body-tracking suit, is implemented using an acting script (presented on Fig. 8.20). During the experimental sessions, the signals which were recorded for analysis included data from: IMPEP 2 time-stamped video and audio logging; camera network capturing several external points of view; body-tracking poses. All signals were synchronised through common-server timestamping.

8.4 Experimental Results

Five experimental sessions were conducted to test the performance of the hierarchical framework presented in this text, in particular to demonstrate its properties of emergence, scalability and adaptivity, as summarised in Table 8.1. Several repetitions of each of these sessions were conducted under roughly the same conditions, so as to confirm reproducibility of the same behaviours. Consequently, in the following lines, the results of each of these sessions will be discussed, and a summary of these findings will be presented in Table 8.2.

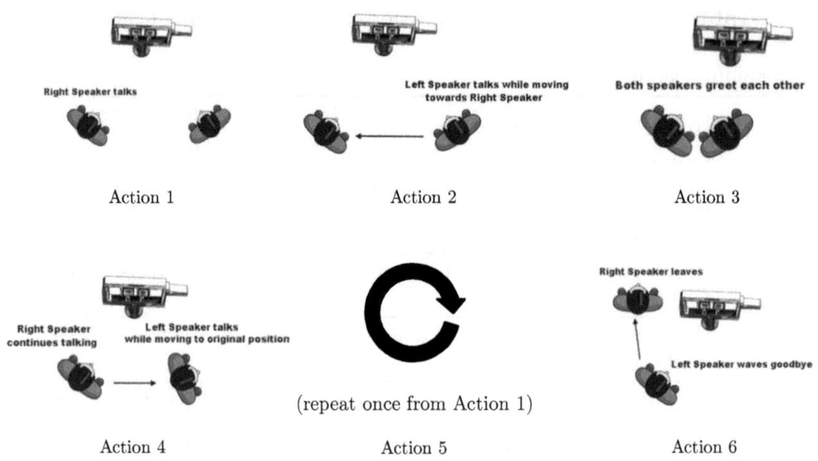

Fig. 8.20. Acting script for active perception experiments

During all the experiments, three views were also filmed from external cameras – see Fig. 8.19 for an overview of the experimental setup using one of these views – and a body-tracking suit was also used by the speaker to the left from the IMPEP head's perspective, the only speaker allowed to walk from one position to another within the BVM horopter (i.e. the portion of spherical volume being perceived and consequently represented by the map), for positioning ground-truth.

Experimental Session 1 – active perception hierarchy implementing all behaviours, using baseline priorities

In this session, a two-speaker scenario was enacted following a script (Fig. 8.20) roughly describing the activity reported in the annotated timeline of the experiment presented on Fig. 8.21.

The genetically imprinted parameters for the distributions that was used was presented on Fig. 8.11. This particular choice of parameters was made to emphasise socially-oriented, high-level behaviours as opposed to low-level behaviours and the IoR effect, which has a noticeable effect only when the former are absent. Countering the IoR effect in the presence of socially-oriented behaviours allows for an apparently more natural emergent behaviour of the system.

The empirical process of finding working parameters within the restrictions described above involved minimal trial-and-error. The greatest restriction was found to be the proportion between weights of auditory saliency and of visual saliency (more specifically, in this case, face detection, the second highest priority). Nevertheless, the range of acceptable proportions was still found to be large enough to be easy to pinpoint.

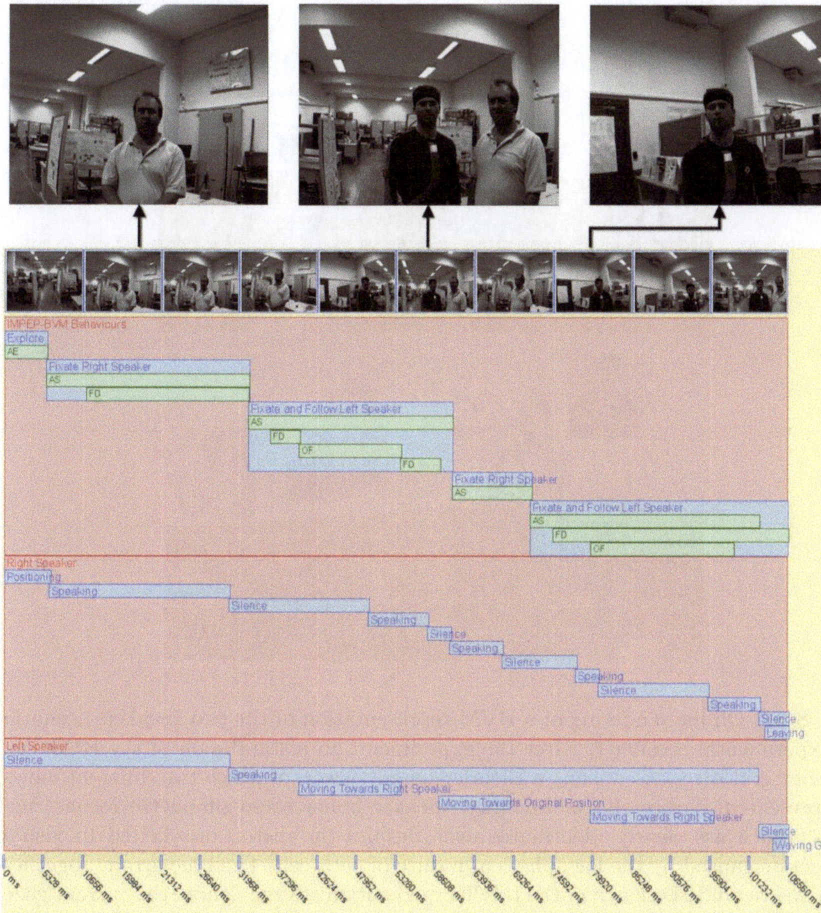

Fig. 8.21. Annotated timeline for Experimental Session 1 – active perception hierarchy implementing all behaviours using baseline priorities. Annotation was performed by three different observers, who were naïve to the underlying basic behaviours and weighting process; the timeline shows the rough average of these annotations, which were very similar (subtle semantic differences between annotation label texts were filtered out, resulting in a maximum temporal difference of annotations of approximately 1 s between annotators). The two lower annotation lanes, labelling the actions performed by the right and left speaker in the perspective of the IMPEP head, were performed by inspection of images taken by the IMPEP stereovision system, by the external cameras, by the tracking suit, and by the audio file recorded by the IMPEP binaural system. The top annotation lane, labelling the emergent behaviours of the active perception system and an interpretation of what were the most prominent underlying low-level behaviours (AE: active exploration; AS: auditory saliency; OF: optical flow magnitude saliency; FD: face detection saliency), was annotated by additionally inspecting saved logs of $P([G^t = C] | V^{1 \to t}\, O^{1 \to t}\, S^t\, G^{t-1}\, \pi_C)$.

Fig. 8.22. Offline rendering of a BVM representation of the two speakers scenario of Experimental Session 1. After the experiment, an instantiation of the BVM occupancy grid is rendered using a Blender-based viewer, of which two different views are presented. Notice the well-defined speaker upper torso silhouette reconstructions, which are clearly identifiable even despite the distortion elicited to visual inspection caused by the log-spherical nature of each cell. The blue arrow, together with an oriented 3D sketch of the IMPEP perception system denote the current gaze orientation. All results depict frontal views, with Z pointing outward. The parameters for the BVM are as follows: $N = 10$, $\rho_{Min} = 1000\,\text{mm}$ and $\rho_{Max} = 2500\,\text{mm}$, $\theta \in [-180°, 180°]$, with $\Delta\theta = 1°$, and $\phi \in [-90°, 90°]$, with $\Delta\phi = 1°$, corresponding to $10 \times 360 \times 180 = 648,000$ cells, approximately delimiting the so-called "personal space" (the zone immediately surrounding the observer's head, generally within arm's reach and slightly beyond, within 2 m range [49]).

As can be seen in Fig. 8.21, the system successfully fixated both speakers, and even exhibited an emergent behaviour very similar to smooth pursuit while following the speaker to the left in the perspective of the IMPEP head. Therefore, a purely saccade-generating model yields, through emergence, a behaviour that closely resembles smooth pursuit. After analysing logs for $P([G^t = C]|V^{1\to t}\, O^{1\to t}\, S^t\, G^{t-1}\, \pi_C)$, it was found that probabilities for saliency moved across the occupancy grid smoothly, given system parameters and temporal performance. This shows that the baseline priority rationale for the choice of parameters for the distributions was reasonably

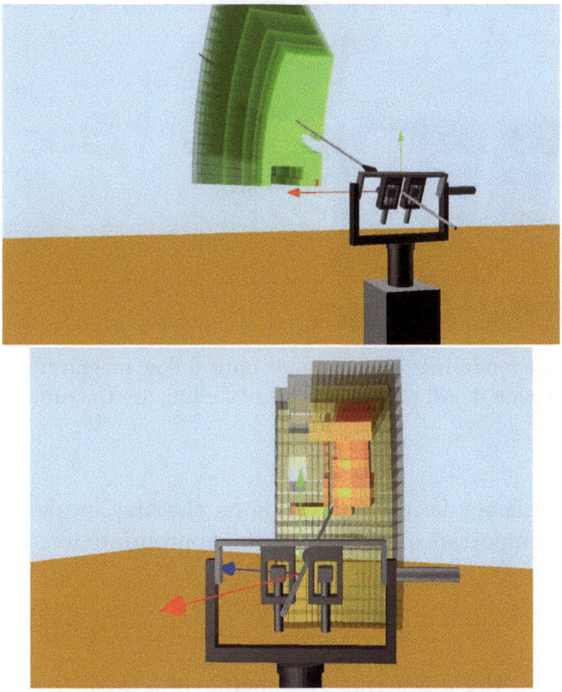

Fig. 8.23. Offline rendering of example saliency maps of the two speakers scenario of Experimental Session 1. The rendering represents values for $P([G^t = C]|V^{1\to t}\,O^{1\to t}\,S^t\,G^{t-1}\,\pi_C)$ that were logged during the session for a specific time instant. Only a slice corresponding to all cells at 10° in azimuth and 20° in elevation around the next fixation point G^t with $P(O_C|C) > .45$ are shown, depicted using a smoothly gradated red-to-green colour-code (red corresponds to lower values, green corresponds to higher values). All other parameters and labelling are the same or analogous to Fig 8.22. On the left, a purely auditory-elicited map is shown, while on the right, a map resulting from the fusion of at least auditory and face detection conspicuity maps is shown.

planned, but, more importantly, clearly demonstrates emergence due to fusion as more than just a pure "sum of parts".

Offline high-definition renderings of BVM and saliency logs are presented on Figs. 8.22 and 8.23, respectively.

Experimental Session 2 – active perception hierarchy implementing all behaviours, with swapped priorities

In this session, the first part of the script of Experimental Session 1 was reenacted, but this time swapping the parameters of the distributions for auditory saliency and face detection saliency, presented on Fig. 8.11. This resulted in the system being unable to change gaze direction to the second speaker after fixating the first speaker, due to the deadlock caused by the

Fig. 8.24. Offline rendering of an example optical flow magnitude saliency map of Experimental Session 4. All parameters and labelling are the same or analogous to Fig 8.22.

face detection saliency keeping attention on the first speaker's face, further showcasing the importance of choosing the appropriate weights for each behaviour.

Experimental Session 3 – active perception hierarchy implementing active exploration only
In this session, the full script of Experimental Session 1 was reenacted, but this time all behaviours except entropy gradient-based active exploration were turned off by making all other distributions uniform. As expected, the behaviour described in Ferreira, Pinho, and Dias [12] emerged, namely the typical "chicken-like" saccadic movements of the IMPEP head exploring the surrounding environment, and a particular sensitivity to the entropy caused by binaural sensing and motion.

Experimental Session 4 – active perception hierarchy implementing optical flow magnitude saliency only
In this session, a single human subject (using the body-tracking suit) is tracked while walking from one position to another within the system's horopter using only optical flow magnitude saliency by making all other distributions uniform, as before. As long as the subject walked within reasonable velocity limits, the system was able to track him successfully.

A saliency map from this session, representing an example of an optical flow magnitude conspicuity map, is presented on Fig. 8.24.

Experimental Session 5 – active perception hierarchy implementing Inhibition of Return only
In this session, the IoR behaviour was tested by making all other distributions uniform, as before. In this case, a fortuitous saccadic behaviour emerged, with the system redirecting gaze to random directions at a constant rate.

Table 8.2. Summary table of experimental session results

Session	Result
1	System performance was annotated by human evaluators, and logs of saliency maps were analysed in comparison, showing that an appropriate choice of weights results in a reasonably human-like emergent behaviour, essential for HRI.
2	System performance was evaluated, as with session 1, proving that modifying the weights for each behaviour change emergence drastically, thus demonstrating the framework's adaptivity.
3, 4 and 5	Removing a specific basic behaviour is shown to change emergent behaviour while maintaining consistency, thus demonstrating the scalability of the system.

8.5 Overall Conclusions and Future Work

In conclusion, the Bayesian hierarchical framework presented in this chapter was shown to adequately follow human-like active perception behaviours, namely by exhibiting the following desirable properties:

Emergence – High-level behaviour results from low-level interaction of simpler building blocks.
Scalability – Seamless integration of additional inputs is allowed by the Bayesian Programming formalism used to state the models of the framework.
Adaptivity – Initial "genetic imprint" of distribution parameters may be changed "on the fly" through parameter manipulation, thus allowing for the implementation of goal-dependent behaviours (i.e. top-down influences).

Future improvements to this framework naturally involve taking advantage of its scalability to include new relevant behaviours, and, most importantly, to capitalise on its adaptivity in order to implement a goal-oriented system in which active perception emergent behaviour changes depending on the task being performed by the robot (Fig. 8.25).

Further future work involves exploiting an important facet of Bayesian frameworks, namely parameter learning – the system will be trained by human subjects using a head-mounted device. The subjects' tracked head-eye gaze shifts control the virtual stereoscopic-binaural point of view, and hence the progression of each stimulus movie – see Fig. 8.26 – while logs of audiovisual stimuli and corresponding fixation points will be logged. This way, controlled free-viewing conditions will be enforced by proposing both generic and specific tasks to the subjects, thus enabling as systematic estimation of distribution parameters in order to construct a lookup table of weights to promote the appropriate human-like emergent behaviour depending on the

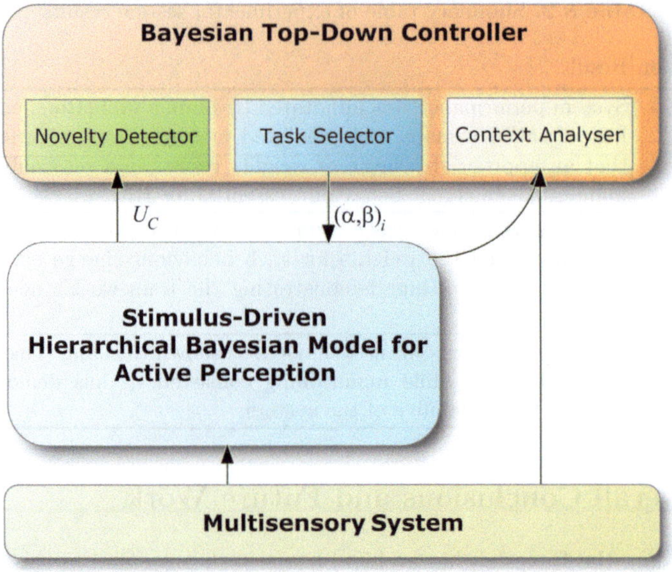

Fig. 8.25. Proposal for goal-oriented active perception framework, including both bottom-up and top-down influences. Corbetta and Shulman [37] posited the hypothesis that a part of the human brain's ventral system acts as a circuit breaker of ongoing cognitive activity when a behaviourally relevant stimulus is detected. According to these authors, when subjects detect an unexpected event, they must break the current attentional set and adopt a new one on the basis of the incoming stimulus. The framework presented in this diagram implements this hypothesis, assuming the set of parameters (α, β) as the "current attentional set", as defined by Corbetta and Shulman [37], with the entropy gradient-based factor U_C for the current time instant being checked for abrupt changes in order for the novelty detector to recognise unexpected events.

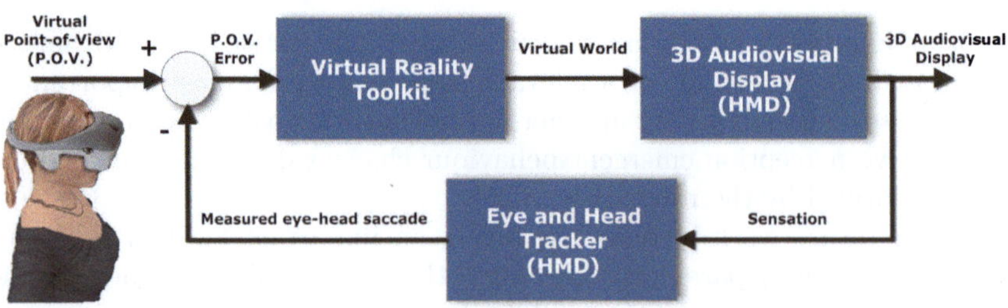

Fig. 8.26. Virtual point-of-view generator setup that allows the updating of audiovisual stimuli presentation according to the monitored subjects' gaze direction

robot's goal. On the other hand, this learning process will allow testing both of our primary hypotheses for active visuoauditory perception, namely active exploration and automatic orienting using sensory saliency, as valid strategies in human behaviour regarding saccade generation. Preliminary work in this respect has already been reported by Ferreira, Tsiourti, and Dias [3].

More details and results concerning the work presented herewith can be found at `http://mrl.isr.uc.pt/projects/BayesianMultimodalPerception`.

References

1. Ferreira, J.F., Lobo, J., Bessiére, P., Castelo-Branco, M., Dias, J.: A Bayesian Framework for Active Artificial Perception. IEEE Transactions on Systems, Man, and Cybernetics, Part B: Cybernetics 43(2), 699–711 (2013) ISSN 1083-4419, doi:10.1109/TSMCB.2012.2214477 185
2. Ferreira, J.F., Castelo-Branco, M., Dias, J.: A hierarchical Bayesian framework for multimodal active perception. Adaptive Behavior 20(3), 172–190 (2012), doi:10.1177/1059712311434662, Published online ahead of print, March 1 185
3. Ferreira, J.F., Tsiourti, C., Dias, J.: Learning emergent behaviours for a hierarchical Bayesian framework for active robotic perception. Cognitive Processing 13(1), 155–159 (2012), ISSN 1612-4782, doi:10.1007/s10339-012-0481-9 223
4. Ferreira, J.F.: Bayesian Cognitive Models for 3D Structure and Motion Multimodal Perception. Ph.D. thesis, Faculty of Sciences and Technology of the University of Coimbra (FCTUC) (2011) (submitted in 2010) 185
5. Ferreira, J.F., Lobo, J., Dias, J.: Bayesian real-time perception algorithms on GPU — Real-time implementation of Bayesian models for multimodal perception using CUDA. Journal of Real-Time Image Processing 6(3), 171–186 (2011) 185, 213
6. Lee, M.D.: How cognitive modeling can benefit from hierarchical Bayesian models. Journal of Mathematical Psychology 55(1), 1–7 (2011); Special Issue on Hierarchical Bayesian Models 187
7. Colas, F., Diard, J., Bessiére, P.: Common Bayesian Models For Common Cognitive Issues. Acta Biotheoretica 58(2-3), 191–216 (2010) 187, 202
8. Ferreira, J.F., Dias, J.: A Bayesian Hierarchical Framework for Multimodal Active Perception. In: "Smarter Sensors, Easier Processing" - Workshop, 11th International Conference on Simulation of Adaptive Behavior, SAB 2010 (2010) 185
9. Bohg, J., Barck-holst, C., Huebner, K., Ralph, M., Rasolzadeh, B., Song, D., Kragic, D.: Towards Grasp-oriented Visual Perception For Humanoid Robots. International Journal of Humanoid Robotics 3(3), 387–434 (2009) 192
10. Colas, F., Flacher, F., Tanner, T., Bessiére, P., Girard, B.: Bayesian models of eye movement selection with retinotopic maps. Biological Cybernetics 100, 203–214 (2009) 199
11. de Croon, G.C.H.E., Sprinkhuizen-Kuyper, I.G., Postma, E.O.: Comparing Active Vision Models. Image and Vision Computing 27(4), 374–384 (2009) 192
12. Ferreira, J.F., Pinho, C., Dias, J.: Implementation and Calibration of a Bayesian Binaural System for 3D Localisation. In: 2008 IEEE International Conference on Robotics and Biomimetics (ROBIO 2008), Bangkok, Thailand (2009) 220

13. Ferreira, J.F., Prado, J., Lobo, J., Dias, J.: Multimodal Active Exploration Using A Bayesian Approach. In: IASTED International Conference in Robotics and Applications, Cambridge MA, USA, pp. 319–326 (2009) 185
14. Gallup, D.: CUDA Stereo (2009), `http://www.cs.unc.edu/~gallup/stereo-demo` 207
15. Hauagge, D.C.: CUDABOF — Bayesian Optical Flow on NVidia's CUDA (2009), `http://www.liv.ic.unicamp.br/ hauagge/Daniel_Cabrini_Hauagge/Home_Page.html` 205
16. Itti, L., Baldi, P.: Bayesian surprise attracts human attention. Vision Research 49, 1295–1306 (2009) 191
17. Watson, T.L., Krekelberg, B.: The Relationship between Saccadic Suppresion and Perceptual Stability. Current Biology 19(12), 1040–1043 (2009) 213
18. Elazary, L., Itti, L.: Interesting objects are visually salient. Journal of Vision 8(3), 1–15 (2008) 190
19. Ferreira, J.F., Bessiére, P., Mekhnacha, K., Lobo, J., Dias, J., Laugier, C.: Bayesian Models for Multimodal Perception of 3D Structure and Motion. In: International Conference on Cognitive Systems (CogSys 2008), pp. 103–108. University of Karlsruhe, Karlsruhe (2008) 194
20. Rommelsea, N.N., der Stigchelc, S.V., Sergeant, J.A.: A review on eye movement studies in childhood and adolescent psychiatry (2008), Article in press; corrected proof, doi:10.1016/j.bandc.2008.08.025 188
21. Castelo-Branco, M., Mendes, M., Sebastião, A.R., Reis, A., Soares, M., Saraiva, J., Bernardes, R., Flores, R., Pérez-Jurado, L., Silva, E.: Visual phenotype in Williams-Beuren syndrome challenges magnocellular theories explaining human neurodevelopmental visual cortical disorders. Journal of Clinical Investigation 117(12), 3720–3729 (2007) 188
22. Dankers, A., Barnes, N., Zelinsky, A.: A Reactive Vision System: Active-Dynamic Saliency. In: 5th International Conference on Computer Vision Systems, Bielefeld, Germany (2007) 192
23. Ferreira, J.F., Castelo-Branco, M.: 3D Structure and Motion Multimodal Perception. State-of-the-Art Report, Institute of Systems and Robotics and Institute of Biomedical Research in Light and Image, University of Coimbra. Bayesian Approach to Cognitive Systems (BACS) European Project (2007) 188, 202
24. Koene, A., Morén, J., Trifa, V., Cheng, G.: Gaze shift reflex in a humanoid active vision system. In: 5th International Conference on Computer Vision Systems, Bielefeld, Germany (2007) 189, 192
25. Lu, Y.C., Christensen, H., Cooke, M.: Active binaural distance estimation for dynamic sources. In: Interspeech 2007, Antwerp, Belgium, pp. 574–577 (2007) 207
26. NVIDIA (2007) CUDA Programming Guide ver 1.2 211
27. Shic, F., Scassellati, B.: A Behavioral Analysis of Computational Models of Visual Attention. International Journal of Computer Vision 73(2), 159–177 (2007) 189, 190, 191
28. Tsotsos, J., Shubina, K.: Attention and Visual Search: Active Robotic Vision Systems that Search. In: The 5th International Conference on Computer Vision Systems, Bielefeld (2007) 192
29. Carmi, R., Itti, L.: Causal saliency effects during natural vision. In: ACM Eye Tracking Research and Applications, pp. 1–9 (2006) 190

30. Dankers, A., Barnes, N., Zelinsky, A.: Active Vision for Road Scene Awareness. In: IEEE Intelligent Vehicles Symposium (IVS 2005), Las Vegas, USA, pp. 187–192 (2005) 192
31. Rocha, R., Dias, J., Carvalho, A.: Cooperative Multi-Robot Systems: a study of Vision-based 3-D Mapping using Information Theory. Robotics and Autonomous Systems 53(3-4), 282–311 (2005) 196
32. Tatler, B.W., Baddeley, R.J., Gilchrist, I.D.: Visual correlates of fixation selection: Effects of scale and time. Vision Research 45(5), 643–659 (2005) 190
33. Carpenter, R.H.S.: The saccadic system: a neurological microcosm. Advances in Clinical Neuroscience and Rehabilitation 4, 6–8 (2004) Review Article 213
34. Caspi, A., Beutter, B.R., Eckstein, M.P.: The time course of visual information accrual guiding eye movement decisions. Proceedings of the National Academy of Sciences USA 101(35), 13086–13090 (2004) 213
35. Ouerhani, N., von Wartburg, R., Hugli, H., Muri, R.: Empirical Validation of the Saliency-based Model of Visual Attention. Electronic Letters on Computer Vision and Image Analysis 3(1), 13–24 (2004) 190
36. Turano, K.A., Geruschat, D.R., Baker, F.H.: Oculomotor strategies for the direction of gaze tested with a real-world activity. Vision Research 43, 333–346 (2003) 190
37. Corbetta, M., Shulman, G.L.: Control of goal-directed and stimulus-driven attention in the brain. Nature Reviews Neuroscience 3, 201–215 (2002), doi:10.1038/nrn755 222
38. Dyde, R.T., Milner, A.D.: Two illusions of perceived orientation: one fools all of the people some of the time; the other fools all of the people all of the time. Experimental Brain Research 144, 518–527 (2002) 187, 188
39. Kopp, L., Gärdenfors, P.: Attention as a minimal criterion of intentionality in robots. Cognitive Science Quarterly 2, 302–319 (2002) 186, 189
40. Parkhurst, D., Law, K., Niebur, E.: Modeling the role of salience in the allocation of overt visual attention. Vision Research 42, 107–123 (2002) 189, 190
41. Breazeal, C., Edsinger, A., Fitzpatrick, P., Scassellati, B.: Active Vision for Sociable Robots. IEEE Transactions on Systems, Man, and Cybernetics–Part A: Systems and Humans 31(5), 443–453 (2001) 192
42. Shibata, T., Vijayakumar, S., Conradt, J., Schaal, S.: Biomimetic Oculomotor Control. Adaptive Behaviour - Special Issue on Biologically Inspired and Biomimetic System 9(3-4), 189–208 (2001) 192
43. Carpenter, R.H.S.: The neural control of looking. Current Biology 10, 291–293 (2000) Primer 213
44. Ballard, D.H.: An Introduction to Natural Computation. MIT Press, Cambridge (1999) 187
45. Breazeal, C., Scassellati, B.: A Context-Dependent Attention System for a Social Robot. In: Sixteenth International Joint Conference on Artificial Intelligence table of contents, pp. 1146–1153 (1999) 190
46. Simoncelli, E.P.: Bayesian Multi-Scale Differential Optical Flow. In: Jähne, B., Haussecker, H., Geissler, P. (eds.) Handbook of Computer Vision and Applications. Academic Press (1999) 205
47. Itti, L., Koch, C., Niebur, E.: A Model of Saliency-Based Visual Attention for Rapid Scene Analysis. IEEE Transactions on Pattern Analysis and Machine Intelligence 20(11), 1254–1259 (1998) 189, 190, 191, 192, 205

48. Murphy, K.J., Carey, D.P., Goodale, M.A.: The Perception of Spatial Relations in a Patient with Visual Form Agnosia. Cognitive Neuropsyshology 15(6/7/8), 705–722 (1998) 187
49. Cutting, J.E., Vishton, P.M.: Perceiving layout and knowing distances: The integration, relative potency, and contextual use of different information about depth. In: Epstein, W., Rogers, S. (eds.) Handbook of Perception and Cognition, vol. 5. Academic Press (1995) Perception of space and motion 199, 218
50. Niebur, E., Itti, L., Koch, C.: Modeling the "where" visual pathway. In: Sejnowski, T.J. (ed.) 2nd Joint Symposium on Neural Computation, Caltech-UCSD. Institute for Neural Computation, La Jolla, vol. 5, pp. 26–35 (1995) 190, 191
51. Tsotsos, J.K., Culhane, S.M., Wai, W.Y.K., Lai, Y., Davis, N., Nuflo, F.: Modeling visual attention via selective tuning. Artificial Intelligence 78, 507–545 (1995) 190
52. Burr, D.C., Morrone, M.C., Ross, J.: Selective suppression of the magnocellular visual pathway during saccadic eye movements. Nature 371, 511–513 (1994) Letters to nature 213
53. Goodale, M.A., Milner, A.D.: Separate visual pathways for perception and action. Trends in Neurosciences 15(1), 20–25 (1992) 187
54. Elfes, A.: Using occupancy grids for mobile robot perception and navigation. IEEE Computer 22(6), 46–57 (1989) 193
55. Aloimonos, J., Weiss, I., Bandyopadhyay, A.: Active Vision. International Journal of Computer Vision 1, 333–356 (1987) 191
56. Bajcsy, R.: Active perception vs passive perception. In: Third IEEE Workshop on Computer Vision, Bellair, Michigan, pp. 55–59 (1985) 191
57. Mishkin, M., Ungerleider, L.G., Macko, K.A.: Object vision and spatial vision: two cortical pathways. Trends in Neuroscience 6, 414–417 (1983) 187
58. Marr, D.: Vision: A Computational Investigation into the Human Representation and Processing of Visual Information. W. H. Freeman and Company, S. Francisco (1982) ISBN-13: 978-0716715672 187
59. Ungerleider, L.G., Mishkin, M.: Two cortical visual systems. In: Ingle, D.J., Goodale, M.A., Mansfield, R.J.W. (eds.) Analysis of Visual Behaviour. MIT Press, Cambridge (1982) 187

9

Wrapping Things Up...

If we will only allow that, as we progress, we remain unsure, we will leave opportunities for alternatives. We will not become enthusiastic for the fact, the knowledge, the absolute truth of the day, but remain always uncertain... In order to make progress, one must leave the door to the unknown ajar.

unsourced quote, credited to Richard Feynman

That man is prudent who neither hopes nor fears anything from the uncertain events of the future.

Mother of Pearl: The Procurator of Judea, *Anatole France (1892)*

9.1 Introduction

After introducing the reader to the set of tools encompassing probabilistic approaches for robotic perception, we are now in the position of coming full circle regarding our introductory consideration of Chapter 1.

In this chapter, we will evaluate the appropriateness of these approaches when applied to modelling cognition and therefore perception, examine the different levels at which this appropriateness might be accepted or challenged, and assess the relevance of these types of computational approaches when comparing to their competition.

We will close the chapter and this book by offering our outlook on the opportunities and challenges faced by those who embrace Bayesian approaches for robotic perception.

9.2 Why Go Bayesian?

9.2.1 The Bayesian Approach and Modelling Cognition

For many years and for several reasons (in most cases, of technical, computational, or even epistemological nature), probabilistic approaches remained outside of the focus of cognitive sciences [7]. However, in the past couple of decades, as confirmed, for example, by Chater, Tenenbaum, and Yuille [7], probabilistic approaches have become very much ubiquitous.

J.F. Ferreira and J. Dias: *Probabilistic Approaches for Robotic Perception*, Springer Tracts in Advanced Robotics 91,
DOI: 10.1007/978-3-319-02006-8_9,

This stems, according to these authors, from the fact that the restrictions on the use of such approaches having been substantially reduced, mainly due to the significant technical advancements in the development of supporting mathematical (theoretical) and computational (implementation) tools.

As a consequence, Bayesian approaches have had a substantial increase of authors advocating their use for modelling cognition – see, for example, Chater et al. [7], or more recently Tenenbaum et al. [3] – but also of sceptical authors criticising their popularity – see, for example, McClelland, Botvinick, Noelle, Plaut, Rogers, Seidenberg, and Smith [5].

Between the two extremes, many researchers have also suggested that probabilistic approaches might be very useful, but only at specific levels and even scales of explanation of cognitive processes. These levels will be introduced in the following section.

9.2.2 Marr's Levels of Probabilistic Explanation

Chater et al. [7], following the taxonomy adopted by many researchers when assessing the appropriateness of applying probabilistic approaches to modelling cognitive processes, suggest that they should be seen in the light of Marr's three levels of computational explanation [9]: the *computational level*, relating to the nature, logic and inputs and outputs of the cognitive problem being solved; the *algorithmic level*, specifying the details of the representations and processes used to solve such problems; and also the *implementational level*, which specifies how these representations and processes are to be realised in practice.

Most authors, including the defenders of probabilistic approaches (e.g., Jacobs and Kruschke [2]) but also their detractors (e.g., McClelland et al. [5]; Colombo and Seriès [1]), generally accept their usefulness in covering the computational level of explanation – this was the basis of our argumentation in the introductory section of Chapter 1 – but are generally either respectively very cautious or completely against accepting the explanatory capabilities of Bayesian modelling in the other two levels. This happens due to the fact that this particular level of analysis is focussed entirely on tackling the nature of the cognitive problem at hand, with no commitment to the actual representations and processes involved, and most of all to the practical realisation of such a cognitive solution [7].

Although the algorithmic and the implementation levels are obviously relevant for robotic perception modelling, affecting, for example, the acceptance of their biological plausibility, they are most certainly not a necessary condition to the development of effective modelling frameworks, as we have seen repeatedly throughout this book[1]. In fact, what Colombo and Seriès [1] refer

[1] Nonetheless, even the scepticism regarding the explanatory power of probabilistic approaches concerning these two levels is being challenged by exciting new research. More on this later on.

to as an "instrumentalist attitude", the border that neuroscience should not cross given the current evidence, is precisely the starting stance that should be taken by the robotics community when assessing the usefulness of these approaches.

Consequently, all that is left in the argument for or against using Bayesian modelling for robotic perception is the assessment of how it compares to its competition.

9.2.3 The Bayesian Approach and Its Competitors

Perception, as a particular manifestation of cognition, may be tackled using conventional logic, non-monotonic logic, heuristic techniques, symbolic rule-based processing, decision trees, connectionist networks (e.g. artificial neural networks), and many others [7; 8]. So, returning to the question that titles this section, "why go Bayesian" indeed? There are many possible answers to this question, some of which are presented next – we hope that the reader finds this analysis enlightening when making a final decision.

First and foremost, Bayesian models inherently and directly deal with uncertainty, namely in what concerns the "irreducible incompleteness of the model", sensor fusion, and latent variables. Secondly, Bayesian approaches allow the introduction of preliminary knowledge, either through priors or through learnt causal relationships. As a consequence, this allows the modeller to access the nature of the problem domain, in fact the explanatory power of these approaches that was introduced previously. Both these genetic advantages were extensively demonstrated throughout the course of this book.

Two additional advantages of Bayesian approaches which are not always readily apparent reside on the strengths of Bayesian learning, as hinted in Chapter 6: it can handle incomplete data sets, but it also offers "an efficient and principled approach for avoiding the overfitting of data" [8], therefore allowing all available data to be used for training, contrary to what happens, for example, with its most direct competitors capable of producing probabilistic outputs, such as probabilistic neural networks.

A final and major advantage, as we have learnt throughout this book, is that it is capable of incorporating inference, decision and learning into a *unifying framework*, as follows (see Bessière et al. [6]; Colas et al. [4] and also Fig. 9.1):

- Firstly, the modeller should take the "irreducible incompleteness of the model" and encode it as uncertainty. This is achieved by introducing whatever preliminary knowledge exists on the problem domain (i.e., priors and structure via π) and by providing the system with the capability of learning all of the missing links (i.e., from experimental data δ, Chapter 6).
- Then, the modeller should endow the framework with the capability of applying the two basic rules of inference, namely the conjunction and the

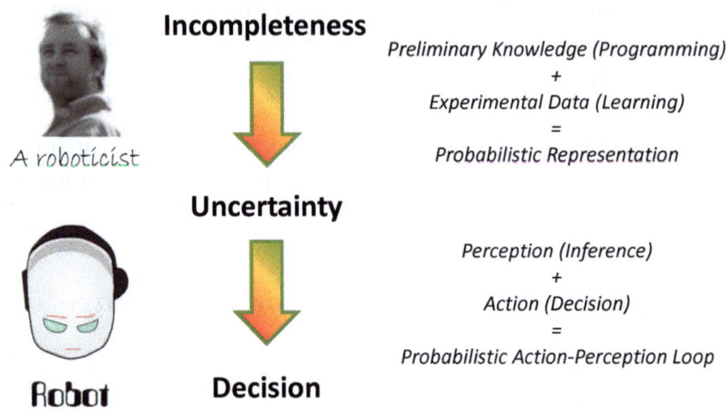

Fig. 9.1. Probabilistic approaches as an unifying framework for robotic perception (adapted from the figures "Probability as an alternative to logic" of [6; 4])

normalisation rules (Chapter 3), and the ability to decide and act on this knowledge (Chapter 5).

Of course, all of this comes with a price: as discussed in Chapter 3, when scaled-up to real-world problems, *full* Bayesian computations are more than often intractable. However, as was also demonstrated in Chapter 3, many technical advances have been achieved in implementing approximate inference, to a point of achieving remarkable precision rates.

On the other hand, in the case of intractability due to the exponential growth of simple Bayesian computations repeated billions of times by the exploitation of independence assumptions and spatial discretisation, such as in the case of the inference grids mentioned throughout the book, an exciting new path has opened due to the recent advances in massive parallel processing.

Therefore, the ubiquitousness of probabilistic approaches for robotic perception is facing a promising future, with which we would like to finish this book, in order to entice the reader for bigger and better things probabilistic.

9.3 The Probabilistic Roadmap – Hopes for the Future

When observing Fig. 9.2, which diagnoses and foresees respectively the current and near future scientific impact of probabilistic approaches for robotic perception, the authors of this book cannot but feel excited by the perspectives open to this field of research.

We feel that a roadmap is unravelling for the following decades, including the following possible routes to explore:

- the accelerating loop of feedback between new, more complex models, and new, more efficient inference tools (Fig. 9.2(b));

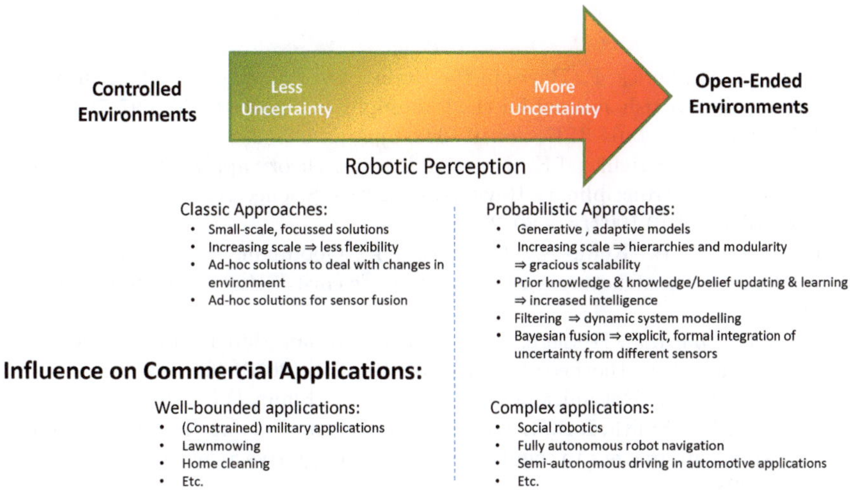

(a) General impact of probabilistic approaches in comparison with competitors

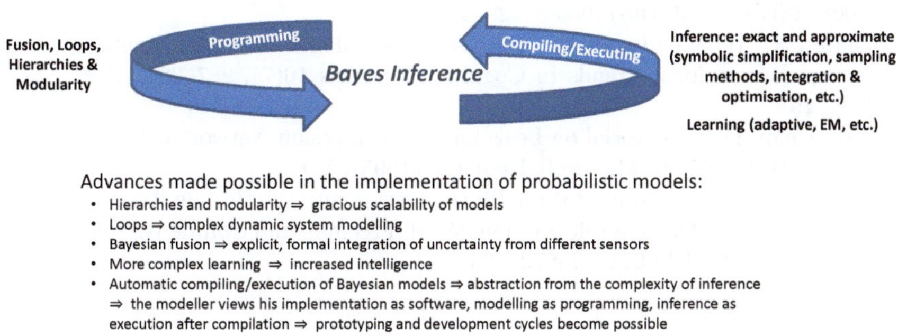

(b) Impact of complex Bayesian modelling methods

Fig. 9.2. Assessment of the impact of probabilistic approaches for robotic perception

- the introduction of new computational technology at the service of these models and inference tools, such as GPUs and reconfigurable hardware;
- the emergence of new processing architectures that might implement plausible logic as an alternative to traditional logic *directly* (in the form, for example, of probabilistic logic gates), potentially resulting in a revolution in low-power processing.

We will certainly trying to be a part of these exploration efforts, and we would like that the final message of this book to be an invitation for the reader to join us.

References

1. Colombo, M., Seriés, P.: Bayes in the Brain – On Bayesian Modelling in Neuroscience. The British Journal for the Philosophy of Science 63(3), 697–723 (2012) ISSN 0007-0882, 1464-3537, doi:10.1093/bjps/axr043 228
2. Jacobs, R.A., Kruschke, J.K.: Bayesian learning theory applied to human cognition. Wiley Interdisciplinary Reviews: Cognitive Science 2(1), 8–21 (2011) ISSN 1939-5086, doi:10.1002/wcs.80 228
3. Tenenbaum, J.B., Kemp, C., Griffiths, T.L., Goodman, N.D.: How to grow a mind: Statistics, structure, and abstraction. Science 331(6022), 1279–1285 (2011) 228
4. Colas, F., Diard, J., Bessiére, P.: Common Bayesian Models For Common Cognitive Issues. Acta Biotheoretica 58(2-3), 191–216 (2010) 229, 230
5. McClelland, J.L., Botvinick, M.M., Noelle, D.C., Plaut, D.C., Rogers, T.T., Seidenberg, M.S., Smith, L.B.: Letting structure emerge: connectionist and dynamical systems approaches to cognition. Trends in Cognitive Sciences 14(8), 348–356 (2010) 228
6. Bessiére, P., Laugier, C., Siegwart, R. (eds.): Probabilistic Reasoning and Decision Making in Sensory-Motor Systems. STAR, vol. 46. Springer, Heidelberg (2008) ISBN 978-3-540-79006-8 229, 230
7. Chater, N., Tenenbaum, J.B., Yuille, A.: Probabilistic models of cognition: Conceptual foundations. Trends in Cognitive Sciences 10(7), 287–291 (2006) 227, 228, 229
8. Heckerman, D.: A Tutorial on Learning With Bayesian Networks. Technical Report MSR-TR-95-06, Microsoft Research (1995) 229
9. Marr, D.: Vision: A Computational Investigation into the Human Representation and Processing of Visual Information. W. H. Freeman and Company, S. Francisco (1982) ISBN-13: 978-0716715672 228

Appendices

A

Introduction to Massive Parallel Programming Using The Compute Unified Device Architecture (CUDA)

A.1 A Brief History of the Implementation of Perception Algorithms Using GPU Computing

GPUs have developed from fixed function architectures to programmable, multi-core architectures, leading to new applications.

A relatively popular subset of this work over the years has been vision and imaging applications. Fung and Mann [4], present an excellent summary on this work, ranging from General Purpose GPU (GPGPU) processing, where graphics hardware is used to perform computations for tasks other than graphics, to the more recent trend of *GPU Computing*, where GPU architectures and programming tools have been developed that have created a parallel programming environment that is no longer based on the graphics processing pipeline, but still exploits the parallel architecture of the GPU — in fact, GPU Computing has transformed the GPGPU concept into the simple mapping of parellelisable algorithms onto SIMD format for the GPU, making a complete abstraction from the intricacies of graphics programming.

As a result, several full-fledged computer vision and image processing toolkits and libraries that resort to GPU technology have emerged, such as OpenVIDIA [10], GPU4Vision [1] or GpuCV [9].

On the other hand, probabilistic approaches to perception have risen the stakes regarding the usefulness of GPU implementations of parallelisable algorithms. Neural network implementation is an example of this, as shown by Jang, Park, and Jung [5], who propose a quick and efficient implementation of neural networks on both GPU and multi-core CPU, with which they developed a text detection system, achieving computational times about 15 times faster than the analogous implementation using CPU and about 4 times faster than implementation on GPU alone.

Occupancy grid-based sensor fusion algorithms, on the other hand, an example of a probabilistic approach to sensor fusion, have as of recently been a source of very interesting work on GPUs, given their obvious parallelisable trait due to the probability independence postulate between grid cells. Moreover, computational frameworks such as this are perfect candidates for GPU processing: very large data structures are processed in parallel using simple operations, yielding the perfect backdrop for SIMD-based computation. However, GPU implemetations for such algorithms are still very recent and few — examples would be the work by Reinbothe, Boubekeur, and Alexa [3], and also Yguel, Aycard, and Laugier [8].

Hence we believe that there is a real contribution to be made in this area, specially now, when GPU Computing has taken such a huge step forward, with the appearance of tools such as NVIDIA's CUDA architecture, which will be summarised in the following section.

A.2 The Compute Unified Device Architecture (CUDA)

We will make a brief presentation of the main features of NVIDIA's CUDA, based on the excellent summary by Hussein, Varshney, and Davis [6]. For a detailed description, refer to [7].

A.2.1 Hardware Architecture

In CUDA terminology, the GPU is called the *device* and the CPU is called the *host*. A CUDA device consists of a set of multicore processors. Each multicore processor is simply referred to as a *multiprocessor*. Cores of a multiprocessor work in a SIMD fashion. All multiprocessors have access to three common memory spaces (globally referred to as *device memory*). They are:

Constant Memory: read-only cached memory space.
Texture Memory: read-only cached memory space that is optimized for texture fetching operations.
Global Memory: read/write non-cached memory

Besides the three memory spaces that are common among all multiprocessors, each multiprocessor has an on chip *shared memory* space that is common among its cores. Furthermore, each core has an exclusive access to a read/write non-cached memory space called *local memory*.

Accessing constant and texture memory spaces is as fast as accessing registers on cache hits. Accessing shared memory is as fast as accessing registers as long as there is no bank conflict. On the other hand, accessing global and local memory spaces is much slower, typically two orders of magnitude slower than floating point multiplication and addition[1].

A.2.2 Execution Model

The execution is based on *threads*. A thread can be viewed as a module, called a *kernel*, that processes a single data element of a data stream. Threads are batched in groups called *blocks*, and can only access shared memory from within their respective blocks. The group of blocks that executes a kernel constitutes one *grid*. Each thread has a three-dimensional index that is unique within its block. Each block in a grid in turn has a unique two dimensional index. Knowing its own index and the index of the block in which it resides, each thread can compute the memory address of a data element to process.

A block of threads can be executed only on a single multiprocessor. However, a single multiprocessor can execute multiple blocks simultaneously by time slicing. Threads in a block can communicate with one another via the shared memory space. They can also use it to share data fetched from global memory. There is no means of synchronization among threads in different blocks. The number of threads within a block that can execute simultaneously is limited by the number of cores in a multiprocessor. A group of threads that execute simultaneously is called a *warp*. Warps of a block are concurrently executed by time slicing.

A.2.3 Optimisation Issues

There are some important considerations that need to be taken into account to obtain good performance on CUDA.

- Effect of Branching: If different threads of a warp take different paths of execution, the different paths are serialized, which reduces parallelism.
- Global Memory Read Coalescing: Global memory reads from different threads in a warp can be coalesced. To be coalesced, the threads have to access data elements in consecutive memory locations. Moreover, addresses of all data elements must follow the memory alignment guidelines. Details are in [7].
- Shared Memory Bank Conflict: Reading from shared memory is as fast as reading from registers unless a bank conflict occurs among threads. Simultaneous accesses to the same bank of shared memory are in most cases serialized.
- Writing to Global Memory: In CUDA, two or more different threads, in the same warp, can write simultaneously to the same address in global memory. The order of writing is not specified, but, one is guaranteed to succeed.

[1] However, the new Fermi GPUs from NVIDIA will have Configurable L1 and Unified L2 caches [2].

References

1. GPU4Vision — Accelerating Computer Vision (2009), http://gpu4vision.icg.tugraz.at/ 235
2. NVIDIA, NVIDIA's Next Generation CUDA™ Compute Architecture: Fermi™. Whitepaper, NVIDIA (2009), Published online http://www.nvidia.com/content/PDF/fermi_white_papers/NVIDIA_Fermi_Compute_Architecture_Whitepaper.pdf 237
3. Reinbothe, C., Boubekeur, T., Alexa, M.: Hybrid Ambient Occlusion. In: Proceedings of the Eurographics Symposium on Rendering (2009) 236
4. Fung, J., Mann, S.: Using Graphics Devices in Reverse: GPU-based Image Processing and Computer Vision. In: IEEE Int'l Conf. on Multimedia and Expo, Hannover, Germany (2008) 235
5. Jang, H., Park, A., Jung, K.: Neural Network Implementation Using CUDA and OpenMP. In: Proceedings of the 2008 Digital Image Computing: Techniques and Applications, pp. 155–161. IEEE Computer Society, Washington, DC (2008) 235
6. Hussein, M., Varshney, A., Davis, L.: On Implementing Graph Cuts on CUDA. In: First Workshop on General Purpose Processing on Graphics Processing Units, Boston, MA (2007) 236
7. NVIDIA, CUDA Programming Guide ver 1.2 (2007) 236, 237
8. Yguel, M., Aycard, O., Laugier, C.: Efficient GPU-based Construction of Occupancy Grids Using several Laser Range-finders. International Journal of Autonomous Vehicles (2007) 236
9. Farrugia, J.P., Horain, P., Guehenneux, E., Alusse, Y.: GpuCV: A framework for image processing acceleration with graphics processors. In: 2006 IEEE International Conference on Multimedia and Expo, pp. 585–588 (2006) 235
10. Fung, J., Mann, S., Aimone, C.: OpenVIDIA: Parallel GPU Computer Vision. In: ACM Multimedia 2005, Singapore, pp. 849–852 (2005) 235

Index

MIX
Papier aus verantwortungsvollen Quellen
Paper from responsible sources
FSC® C105338

If you have any concerns about our products, you can contact us on
ProductSafety@springernature.com

In case Publisher is established outside the EU, the EU authorized representative is:
Springer Nature Customer Service Center GmbH
Europaplatz 3, 69115 Heidelberg, Germany

Printed by Libri Plureos GmbH
in Hamburg, Germany